FOREVER UNDECIDED

FOREVER UNDECIDED

A Puzzle Guide
to Gödel

by
Raymond Smullyan

ALFRED A. KNOPF · NEW YORK

1987

THIS IS A BORZOI BOOK
PUBLISHED BY ALFRED A. KNOPF, INC.

Copyright © 1987 by Raymond Smullyan

Library of Congress Cataloging-in-Publication-Data

Smullyan, Raymond.
Forever undecided, a puzzle guide to Gödel.

1. Gödel's theorem.   2. Mathematical recreations.   I. Title.
QA9.65.S68   1987   511.3   86-45297
ISBN 0-394-54943-0

Manufactured in the United States of America
FIRST EDITION

*Once again, I would like to thank my editor, Ann Close,
and the production editor, Melvin Rosenthal, for all their
expert assistance, and my typist, Nancy Spencer, for her
excellent secretarial help.*

Dedicated to All Consistent Reasoners
Who Can Never Know That They Are Consistent

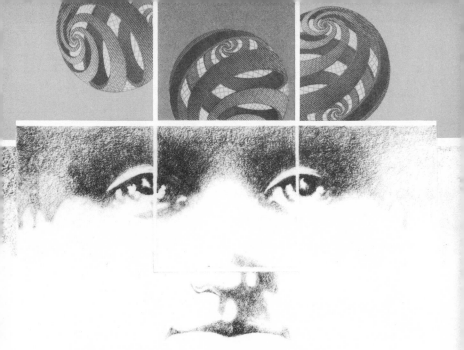

ALTERNATE SELECTION

FOREVER UNDECIDED

A PUZZLE GUIDE TO GÖDEL

"Is it possible for a rational human being to be in a position in which he cannot believe that he is consistent without losing his consistency in the process?"

The search for the answer to this question (modelled on Kurt Gödel's so-called Second Incompleteness Theorem) in Raymond Smullyan's latest and most challenging work leads to heady adventures among knights, knaves and reasoners of diverse postures (conceited, timid, peculiar, among others). Along the way we come to understand Gödel's two great incompleteness theorems, some of their implications, and why we, like Gödel himself, must remain "FOREVER UNDECIDED."

BY RAYMOND SMULLYAN

Author of *To Mock a Mockingbird*, *The Lady or the Tiger?*, and *This Book Needs No Title*

Another captivating puzzle guide from the beguiling Raymond Smullyan—this one conjuring with Gödel's two incompleteness theorems, symbolic logic, self-fulfilling beliefs, and possible world semantics

FOREVER UNDECIDED continues Raymond Smullyan's series of wonderfully engaging puzzle books by transferring Kurt Gödel's famous incompleteness theorems from the formal domain of mathematical systems and the propositions provable in them to the realm of human beings and the propositions believed by them. To this end Smullyan transports us to a magical island where knights tell only the truth and knaves always lie. As warm-ups we encounter a census-taker who has to determine which inhabitants are knights and which are knaves, and a philosopher-logician who can determine the whereabouts of his flighty bird-wife only by questioning these same natives.

Interspersed with these puzzles is an introductory account of symbolic logic and an explanation of how this valuable technique can systematically solve whole tracts of puzzles of the knight-knave variety. Of equal interest to artificial intelligence enthusiasts are Smullyan's puzzles related to self-fulfilling beliefs. Under what circumstances can the mere belief in a proposition cause the belief to be true? Later chapters relate these belief systems to important systems of mathematics and the philosophically fascinating subject of possible world semantics—a field conceived by Leibniz, brought to perfection by Kripke, and of considerable interest in computer science and artificial intelligence today.

Ingenious, instructive, entertaining, FOREVER UNDECIDED shows how a sufficiently wild imagination can tame even the most abstruse mathematical concepts.

"What characterizes [Smullyan's] books and papers," George Boolos, professor of philosophy at MIT, has written, "is how astoundingly clear they are. He is able to strip away what is inessential and give you the core of an idea, undiluted and unvarnished. He is a great simplifier."

THE AUTHOR

In his many marvelously beguiling books, mathematician and logician RAYMOND SMULLYAN has almost single-handedly invented a new genre of puzzle book—one which demonstrates how logic, philosophy, psychology, artificial intelligence, computer science, and mathematics are drawing closer and closer together. He is currently Oscar Ewing Professor of Philosophy at Indiana University and Professor Emeritus of The City University of New York—Lehman College and Graduate Center.

FOREVER Undecided
A Puzzle Guide to Gödel
Raymond Smullyan

- 257 pages
- Published by Alfred A. Knopf in March 1987
(48750)
Publisher's Price $17.95
Member's Price $13.75
Earns one bonus credit toward the four required for your next Bonus Certificate

Contents

CONTENTS

Part IV
LET'S BE CAREFUL!

Part V
THE CONSISTENCY PREDICAMENT

Part VI
SELF-FULFILLING BELIEFS AND LÖB'S THEOREM

Part VII
IN DEEPER WATERS

CONTENTS

Preface

Is it possible for a rational human being to be in a position in which he cannot believe that he is consistent without losing his consistency in the process? That is one of the main themes of this book. It is modelled on the famous discovery of Kurt Gödel (the so-called Second Incompleteness Theorem) that any consistent mathematical system with enough power to do what is known as elementary arithmetic must suffer from the surprising limitation that it can never prove its own consistency!

There are several reasons why I have transferred Gödel's argument from the formal domain of mathematical systems and the propositions provable in them to the realm of human beings and the propositions *believed* by them. For one thing, human beings and their beliefs are far more familiar to the nonspecialist than abstract mathematical systems, and so I can thus explain the essentials of Gödel's ideas in a language that everybody can understand. Also, putting these matters in human terms has an enormous psychological appeal and turns out to be highly relevant to the ever-growing field of artificial intelligence.

As with my earlier puzzle books, I start off with a host of puzzles about liars and truth tellers (knaves and knights). Aside from the novelty of these problems (almost all of them are published here for the first time), there is an important added feature—namely, that interspersed with these puzzles is an introductory account of *symbolic logic* (a subject that is fortunately being taught today in many of our high schools) and an explanation of how this valuable technique can systematically solve whole tracts of puzzles of the knight–knave variety. (This should be of particular interest to teachers of logic—at either the high school or college level—who wish to enliven their logic courses with creative and challenging puzzles.) Therefore, the knowledge that the reader will gain will pay off handsomely in clarifying the later and more profound chapters on Gödel's work.

Another theme of the later chapters—and of equal interest to artificial intelligence—is that of *self-fulfilling beliefs*. Under what circumstances can the mere belief in a proposition cause the belief to be true? (Could this have anything to do with religion?) There is an important theorem of the logician M. H. Löb that is highly relevant here. This theorem is closely related to Gödel's theorem, and when stated first in terms of reasoners and their beliefs, it is within easy grasp of the general reader.

Although my primary emphasis is on belief systems, by no means do I stop there. As the book advances, the reader is led to a deeper and deeper understanding of how these belief systems are related to important systems of mathematics. This in turn leads us to the philosophically fascinating subject of *possible world semantics*—a field conceived by Leibniz and brought to perfection in this century by the logician Saul Kripke. This subject also is playing a major role today in computer science and artificial intelligence.

I have lectured a good deal on all this material to such diverse groups as bright high school students and Ph.D.'s in mathematics, philosophy, and computer science. The responses of both groups were equally gratifying—they were intrigued. Indeed, any neophyte who is good at math or science can thoroughly master this entire book (though some application will be needed), yet many an expert will find here a wealth of completely new and fresh material.

It is remarkable how logic, philosophy, psychology, artificial intelligence, computer science, and mathematics are drawing closer and closer together these days. We are living in an exciting era!

Elka Park, N.Y.　　　　　RAYMOND SMULLYAN
July 1986

· *Part I* ·

YOU MIGHT BE SURPRISED!

· 1 ·

A Diabolical Puzzle

I BELIEVE that the following puzzle may well be the most diabolical puzzle ever invented (and if it is, I proudly take credit for the invention).

Two people—A and B—each make an offer, which is given below. The problem is to determine which of the offers is better.

A's Offer: You are to make a statement. If the statement is true, you get exactly ten dollars. If the statement is false, then you get either less than ten or more than ten dollars, but not exactly ten dollars.

B's Offer: You are to make a statement. Regardless of whether the statement is true or false, you get more than ten dollars.

Which of the two offers would you prefer? Most people decide that B's offer is better, since it *guarantees* more than ten dollars, whereas with A's offer, there is no certainty of winning more than ten. And it does *seem* that B's offer is better, but seeming is sometimes deceptive. In fact, I will make an offer of my own: If any of you are willing to make me A's offer, I'll pay you twenty dollars in advance. Anyone game? (Before taking me up on it, you had better read the rest of this chapter!)

· · ·

I will not give the solution to the above problem until after first considering some simpler, related puzzles.

In my book *To Mock a Mockingbird*, I presented the following puzzle: Suppose I offer two prizes—Prize 1 and Prize 2. You are to make a statement. If the statement is true, then I am to give you one of the two prizes (not saying which one). If your statement is false, then you get no prize. Obviously you can be sure of winning one of the two prizes by saying: "Two plus two is four," but suppose you have your heart set on Prize 1; what statement could you make that would guarantee that you will get Prize 1?

The solution I gave is that you say: "You will not give me Prize 2." If the statement is false, then what it says is not the case, which means that I *will* give you Prize 2. But I can't give you a prize for making a false statement, and so the statement can't be false. Therefore, it must be true. Since it is true, then what it says *is* the case, which means that you will not get Prize 2. But since your statement was true, I must give you one of the two prizes, and since it is not Prize 2, it must be Prize 1.

As we will see shortly, this little puzzle is closely related to Gödel's famous Incompleteness Theorem. To find out how, let us consider a similar puzzle (actually, the same puzzle in a different guise). First, we go to the Island of Knights and Knaves (which plays a prominent role in this book) in which every inhabitant is either a knight or a knave. On this island knights make only true statements and knaves make only false ones.

Suppose now that there are two clubs on this island—Club I and Club II. Only knights are allowed to be members of either club; knaves are rigorously excluded from both. Also, every knight is a member of one and only one of the two clubs. You visit the island one day and meet an unknown native who makes a statement from which you can deduce that he must be a member of Club I. What statement can he have made from which you can deduce this?

In analogy with the last problem, the speaker could have said: "I

4

am not a member of Club II." If the speaker were a knave, then he really couldn't be a member of Club II and would have made a true statement, which a knave cannot do. Therefore he must be a knight, and his statement must have been true; he really isn't a member of Club II. But since he is a knight, he must be a member of a club —so he belongs to Club I.

The analogy should be obvious: Club I corresponds to those who make true statements and will get Prize 1; Club II corresponds to those who make true statements and will get Prize 2.

These puzzles embody the essential idea behind Gödel's famous sentence that asserts its nonprovability in a given mathematical system. Suppose we classify all the *true* sentences of the system (like the knights in the puzzle above) into two groups: Group I consists of all sentences of the system which, though true, are not *provable* in the system; and Group II consists of all the sentences which are not only true but are actually provable in the system. What Gödel did was to construct a sentence that asserted that it was in Group II—the sentence can be paraphrased, "I am not provable in the system." If the sentence were false, then what it says would not be the case, which would mean that it *is* provable in the system, which it cannot be (since all sentences provable in the system are true). Hence the sentence must be true, and, as it says, it is not provable in the system. Thus Gödel's sentence is true, but not provable in the system.

We will have much to say about Gödel sentences in the course of this book. For now, I wish to consider some more variants of the Prize Puzzle.

1 · First Variant

Again I offer two prizes—Prize 1 and Prize 2. If you make a true statement, I will give you at least one of the two prizes and possibly both. If you make a false statement, you get no prize. Suppose you are ambitious and wish to win both prizes. What statement would you make? (Solutions appear at the ends of chapters unless otherwise specified.)

2 · Second Variant

This time the rules change a bit: If you make a true statement, you get Prize 2; if you make a false statement, you don't get Prize 2 (you might or might not get Prize 1). Now what statement will win you Prize 1?

3 · A Perverse Variant

Suppose that in a perverse frame of mind, I now tell you that if you make a false statement, you will get one of the two prizes, but if you make a true statement, you get no prize. What statement will win Prize 1?

4 · More on the Diabolical Puzzle

Now let us return to the puzzle that opened this chapter: What was diabolical was *my* offer to pay you twenty dollars for making me A's offer, because if you took me up on it, I could win as much from you as I liked—say, a million dollars. Can you see how?

SOLUTIONS

1 · A statement that works is: "I will either get both prizes or no prize." If the statement is false, then what it says is not the case, which means that you will get exactly one prize. But you can't get a prize for a false statement. Therefore the statement must be true, and you really will get either both prizes or no prize. Since you did not make a false statement, which would result in no prize, you must get both prizes.

6

2 · Just say: "I will get no prize." If the statement is true, then on the one hand you get no prize (as the statement says), but on the other hand, you get Prize 2 for having made a true statement. This is a clear contradiction, hence the statement must be false. Then, unlike what the statement says, you must get a prize. You can't get Prize 2 for a false statement, so you get Prize 1.

3 · This time say: "I will get Prize 2." I leave the proof to the reader.

4 · All I have to do is say: "You will neither pay me exactly ten dollars nor exactly one million dollars." If my statement is true, then on the one hand you will not pay me either exactly ten dollars or exactly a million dollars, but on the other hand you *must* pay me exactly ten dollars for having made a true statement. This is a contradiction, hence the statement can't be true and must be false. Since it is false, what it says is not the case, which means that you *will* pay me either exactly ten dollars or exactly a million dollars, but you can't pay me exactly ten dollars for a false statement, hence you must pay me a million dollars.

Anyone still game?

· 2 ·

Surprised?

LET US now turn to the paradox at the surprise examination—a paradox that is quite relevant to the consideration of Gödel in this book. We will look at it in the following form: On a Monday morning, a professor said to his class, "I will give you a surprise examination someday this week. It may be today, tomorrow, Wednesday, Thursday, or Friday at the latest. On the morning of the day of the examination, when you come to class, you will not know that this is the day of the examination."

Well, a logic student reasoned as follows: "Obviously I can't get the exam on the last day, Friday, because if I haven't gotten the exam by the end of Thursday's class, then on Friday morning I'll know that this is the day, and the exam won't be a surprise. This rules out Friday, so I now know that Thursday is the last possible day. And, if I don't get the exam by the end of Wednesday, then I'll know on Thursday morning that this must be the day (because I have already ruled out Friday), hence it won't be a surprise. So Thursday is also ruled out."

The student then ruled out Wednesday by the same argument, then Tuesday, and finally Monday, the day on which the professor was speaking. He concluded: "Therefore I can't get the exam at all; the professor cannot possibly fulfill his statement." Just then, the professor said: "Now I will give you your exam." The student was most surprised!

What was wrong with the student's reasoning?

8

Dozens of articles have been written about this famous problem, and uniform agreement about the correct analysis does not yet seem to have been reached. My own view is very briefly as follows:

Let me put myself in the student's place. I claim that I could get a surprise examination on any day, *even on Friday!* Here is my reasoning: Suppose Friday morning comes and I haven't gotten the exam yet. What would I then believe? Assuming I believed the professor in the first place (and this assumption is necessary for the problem), could I consistently continue to believe the professor on Friday morning if I hadn't gotten the exam yet? I don't see how I could. I could certainly believe that I would get the exam today (Friday), but I couldn't believe that I'd get a *surprise* exam today. Therefore, how could I trust the professor's accuracy? Having doubts about the professor, I wouldn't know what to believe. Anything could happen as far as I'm concerned, and so it might well be that I *could* be surprised by getting the exam on Friday.

Actually, the professor said two things: (1) You will get an exam someday this week; (2) You won't know on the morning of the exam that this is the day. I believe it is important that these two statements should be separated. It *could* be that the professor was right in the first statement and wrong in the second. On Friday morning, I couldn't consistently believe that the professor was right about both statements, but I could consistently believe his first statement. However, if I do, then his second statement is wrong (since I will then believe that I *will* get the exam today). On the other hand, if I doubt the professor's first statement, then I won't know whether or not I'll get the exam today, which means that the professor's second statement is fulfilled (assuming he keeps his word and gives me the exam). So the surprising thing is that the professor's second statement is true or false depending respectively on whether I do not or do believe his first statement. Thus the one and only way the professor can be right is if I have doubts about him; if I doubt him, that makes him right, whereas if I fully trust him, that makes him wrong! I don't know whether this curious point has been taken into consideration before.

The One-Day Version. The complication of having more than one day is really irrelevant to the heart of the problem, hence the following "one-day" version of the paradox has been proposed: The professor says to the student: "I will give you a surprise examination today." What is the student to make of that?

An equivalent problem is this: Suppose a student asks his theology professor, "Does God really exist?" The professor answers, "God exists, but you will never believe that God exists." What is the student to make of that? We will see later that under certain very reasonable assumptions about the student's reasoning ability, he cannot believe the professor without becoming inconsistent.

A problem even more pertinent to the heart of this book is this: Again, the student asks his theology professor whether God exists. This time the professor gives the following curious answer: "God exists if and only if you never believe that God exists." Equivalently, the professor is saying: "If God exists, then you will never believe that God exists, but if God doesn't exist, then you will believe that God does exist." What is the poor student to make of *that?* Can the student believe the professor without becoming inconsistent? Yes; it turns out that he can, but (again under certain reasonable assumptions about the student's reasoning ability, which will be explained in the course of this book) the student, who believes the professor, can remain consistent only if he does not *know* that he is consistent! In other words, if the student believes the professor and also believes in his own consistency, then he will become inconsistent.

This paradox is closely related to Gödel's remarkable Second Incompleteness Theorem—the theorem on the nonprovability of consistency. Gödel considered some of the most comprehensive mathematical systems known to this day. These systems are certainly consistent (since everything provable in them is true), but the amaz-

10

ing thing is that these systems, despite their power (or, looked at another way, because of their power), are unable to prove their own consistency. We know the consistency of these systems only by methods that cannot be formalized in the systems themselves.

This book will investigate these paradoxes and Gödel's work. To help us in this pursuit (and to have some fun!), let's turn first to some logic puzzles and their relation to propositional logic.

Part II

THE LOGIC OF LYING
AND TRUTH TELLING

· 3 ·

The Census
Taker

MUCH OF the action in this book will take place on the Island of
Knights and Knaves, where, as we have seen, knights always make
true statements, knaves always make false statements, and every
inhabitant is either a knight or a knave.

A fundamental fact about this island is that it is impossible for any
inhabitant to claim to be a knave, because a knight would never lie
and say he is a knave, and a knave would never truthfully admit to
being a knave.

The following four problems will introduce the logical connectives
and, or, if-then, and *if-and-only-if,* which will be dealt with more
formally in Chapter 6.

McGREGOR'S VISIT

The census taker Mr. McGregor once did some fieldwork on the
Island of Knights and Knaves. On this island, women are also called
knights and knaves. McGregor decided on this visit to interview
married couples only.

1 · (And)

McGregor knocked on one door; the husband partly opened it and
asked McGregor his business. "I am a census taker," replied Mc-

Gregor, "and I need information about you and your wife. Which, if either, is a knight, and which, if either, is a knave?"

"We are both knaves!" said the husband angrily as he slammed the door.

What type is the husband and what type is the wife? (Solution follows Problem 2.)

2 · (Or)

At the next house, McGregor asked the husband: "Are both of you knaves?" The husband replied: "At least one of us is."

What type is each?

Solution to Problem 1. If the husband were a knight, he would never have claimed that he and his wife were both knaves. Therefore he must be a knave. Since he is a knave, his statement is false; so they are not both knaves. This means his wife must be a knight. Therefore he is a knave and she is a knight.

Solution to Problem 2. If the husband were a knave, then it would be true that at least one of the two is a knave, hence a knave would have made a true statement, which cannot be. Therefore the husband must be a knight. It then follows that his statement was true, which means that either he or his wife is a knave. Since he isn't a knave, then his wife is. And so the answer is the opposite of Problem 1—he is a knight and she is a knave.

The next problem is more startling than the preceding two (at least to those who haven't seen it before). It contains a theme that runs through some of the advanced problems that will crop up in later chapters.

3 · (If-Then)

The next home visited by McGregor proved more of a puzzler. The door was opened timidly by a rather shy man. After McGregor asked

16

him to say something about himself and his wife, all the husband said was: "If I am a knight, then so is my wife."

McGregor walked away none too pleased. "How can I tell anything about either of them from such a noncommittal response?" he thought. He was about to write down "Husband and wife both unknown," when he suddenly recalled an old logic lesson from his Oxford undergraduate days. "Of course," he realized, "I can tell *both* their types!"

What type is the husband and what type is the wife?

Solution. Suppose the husband is a knight. Then it is true what he said—namely, that if he is a knight, so is his wife—and hence his wife must also be a knight. This proves that *if* the husband is a knight, so is his wife. Well, that's exactly what the husband said; he said that *if* he is a knight, then so is his wife. Therefore he made a true statement and so he must be a knight. We now know that he is a knight, and we have already proved that if he is a knight, so is his wife. Therefore the husband and wife must both be knights.

The idea behind the last problem has more far-reaching ramifications than the reader might realize. Let us consider the following variant of the problem: Suppose you visit the island prospecting for gold. Before you start digging, you want to find out whether there really is any gold on the island. It is to be assumed that each native knows whether there is any gold on the island or not. Suppose a native says to you: "If I am a knight, then there is gold on the island." You can then justifiably conclude that the native must be a knight and that there must be gold on the island. The reasoning is the same as that of the solution to Problem 3: Suppose the native is a knight, then it is really true that if he is a knight, there is gold on the island and hence that there is gold on the island. This proves that if he is a knight, then there is gold on the island. Since he said just that, he is a knight. Hence there is gold on the island.

The solution to Problem 3 and its variant are special cases of the following fact, which is sufficiently important to record as Theorem I.

Theorem I. Given any proposition p, suppose a native of the knight-knave island says: "If I am a knight, then p." Then the native must be a knight and p must be true.

The solution to Problem 3 is a special case of Theorem I, taking p to be the proposition that the native's wife is a knight. The variant of Problem 3 (the problem about the gold) is also a special case of Theorem I, taking p to be the proposition that there is gold on the island.

It also follows from Theorem I that no inhabitant of the knight-knave island can say: "If I'm a knight, then Santa Claus exists" (unless, of course, Santa Claus really does exist).

4 · (If-and-Only-If)

When the census taker interviewed the fourth couple, the husband said: "My wife and I are of the same type; we are either both knights or both knaves."

(The husband could have alternatively said: "I am a knight if and only if my wife is a knight." It comes to the same thing.)

What can be deduced about the husband and what can be deduced about the wife?

Solution. It cannot be determined whether the husband is a knight or a knave, but the wife's type can be determined as follows:

If the wife were a knave, the husband could never claim that he is the same type as his wife, because that would be tantamount to claiming that he is a knave, which he cannot do.

An alternative way of looking at the problem is this: The husband is either a knight or a knave. If he is a knight, his statement is true, hence he and his wife really are of the same type, which means his wife is also a knight. On the other hand, if he is a knave,

18

then his statement is false, hence he and his wife are of different types, which means that his wife, unlike her husband, is a knight. And so regardless of whether the husband is a knight or a knave, his wife must be a knight. (The husband's type here is "indeterminate"; he could be a knight who truthfully claimed to be like his wife, or he could be a knave who falsely claimed to be like his wife.)

This problem is a special case of the following: Given any proposition p, suppose an inhabitant of the island says, "I am a knight if and only if p is true." What can be deduced?

Two propositions are called *equivalent* if they are either both true or both false—in other words, if either one of them is true, so is the other. Two propositions are called *inequivalent* if they are not equivalent—in other words, if one of them is true and the other is false. Now, the inhabitant has said: "I am a knight if and only if p is true." If we let k be the proposition that the inhabitant is a knight, then the inhabitant is claiming that k is equivalent to p. If he is a knight, then his claim is true, hence k really is equivalent to p; and since k is true (he is a knight), then p is also true. On the other hand, if he is a knave, then his claim is false; k is not really equivalent to p, but since k is false (he is not a knight), then again p must be true (because any proposition inequivalent to a false proposition is obviously true). And so we see that p must be true, but k is indeterminate. Let us record this as Theorem II.

Theorem II. Given any proposition p, suppose an inhabitant says: "I am a knight if and only if p." Then p must be true, regardless of whether the inhabitant is a knight or a knave.

Let us return to the problem of whether there is gold on the island. Suppose a native says: "I am a knight if and only if there is gold on the island." Then according to Theorem II (taking p to be the proposition that there is gold on the island), we see that there

19

must be gold on the island, though we cannot determine whether the native is a knight or a knave.

We see, therefore, that if a native says: "If I am a knight, then there is gold on the island," according to Theorem I, we can infer *both* that he is a knight and that there is gold on the island. But if he instead says: "I am a knight if and only if there is gold on the island," then, according to Theorem II, all we can infer is that there is gold; we cannot determine whether the speaker is a knight or a knave.

Theorem II is the basis of a famous puzzle invented by the philosopher Nelson Goodman. The puzzle can be rendered as follows: Suppose you go to the Island of Knights and Knaves and wish to find out whether or not there is gold on the island. You meet a native and you are allowed to ask him only one question, which must be answerable by yes or no. What question would you ask him?

A question that works is: "Is it the case that you are a knight if and only if there is gold on the island?" If he answers yes, then according to Theorem II there is gold on the island. If he answers no, then there is no gold on the island (because he is denying that his being a knight is equivalent to there being gold on the island), which is tantamount to claiming that his being a knight is equivalent to there *not* being gold on the island, so again according to Theorem II there is no gold on the island.

SOME RELATED PROBLEMS

5

What statement could a native make from which you could deduce that if he is a knight, then there is gold on the island, but if he is a knave, then there might or might not be gold on the island?

6

What statement could a native make such that you could deduce that if there is gold on the island, then he must be a knight, but if there is no gold on the island, then he could be either a knight or a knave?

7

I once visited this island and asked a native: "Is there gold on this island?" All he said in reply was: "I have never claimed that there is gold on this island." Later, I found out that there *was* gold on the island. Was the native a knight or a knave?

SOLUTIONS

5 · There are many statements that would work. One such statement is: "I am a knight and there is gold on the island." Another is: "There is gold on the island and there is silver on the island." (If the native is a knight, then of course there is gold—as well as silver—but if there is gold, the native needn't be a knight—there might not be any silver.)

6 · One statement that works is: "Either I am a knight or there is gold on the island." The phrase "either–or" *means at least one—and possibly both.* And so if there is gold on the island, then it is certainly true that *either* the native is a knight *or* there is gold on the island. Therefore, if there is gold on the island, then the native's statement was true, which in turn implies that the native must be a knight. This proves that if there is gold on the island, then the native is a knight.

On the other hand, the native could be a knight without there being any gold on the island, because if he is a knight then it is true that *either* he is a knight *or* there is gold on the island.

Another statement that works is: "Either there is gold on the island or silver on the island."

7 · Suppose the native were a knave. Then his statement is false, which means that once he *did* claim that there is gold on the island. His claim must have been false (since he is a knave), which means that there is no gold on the island. But I told you that there *was* gold on the island. Hence he can't be a knave; he must be a knight.

· 4 ·

In Search of Oona

THERE IS a whole cluster of knight-knave islands in the South Pacific on which some of the inhabitants are half human and half bird. These bird-people fly just as well as birds and speak as fluently as humans.

This is the story of a philosopher—a logician, in fact—who visited this cluster of islands and fell in love with a bird-girl named Oona. They were married. His marriage was a happy one, except that his wife was too flighty! For example, he would come home at night for dinner, but if it was a particularly lovely evening, Oona would have flown off to another island. So he would have to paddle around in his canoe from one island to another until he found Oona and brought her home. Whenever Oona landed on an island, the natives would all see her in the air and know she was landing. Once she was down, however, it would be very difficult to find her, so the first thing the husband did when he arrived on an island was to try to find out from the natives whether Oona had landed. What made it so difficult, of course, was that some of the natives were knaves and wouldn't tell the truth. Here are some of the incidents that befell him.

1

On one occasion, the husband came to an island in search of Oona and met two natives, A and B. He asked them whether Oona had landed on the island. He got the following responses:

A: If B and I are both knights, then Oona is on this island.
B: If A and I are both knights, then Oona is on this island.

Is Oona on this island?

2

On another occasion, two natives A and B made the following statements:

A: If either of us is a knight, then Oona is on this island.
B: That is true.

Is Oona on this island?

3

I do not remember the details of the next incident too clearly. I know that the logician met two natives A and B and that A said: "B is a knight, and Oona is on this island." But I do not remember exactly what B said. He either said: "A is a knave, and Oona is not on this island." Or he said: "A is a knave, and Oona *is* on this island." I wish I could remember which! At any rate, I do remember that the logician was able to determine whether or not Oona was on the island. Was she?

4

In the next incident, the logician came to a very small island with only six inhabitants. He interrogated each one, and curiously enough, they all said the same thing: "At least one knave on this island has seen Oona land here this evening."

Did any native of this island see Oona land there that evening?

5

In another curious incident, when the husband arrived on an island looking for Oona, he met five natives A, B, C, D, and E, who all guessed his purpose and grinned at meeting him. They made the following statements:

A: Oona is on this island.
B: Oona is *not* on this island!
C: Oona was here yesterday.
D: Oona is not here today, and she was not here yesterday.
E: Either D is a knave or C is a knight.

The logician thought about this for a while, but could get nowhere.

"Won't one of you please make another statement?" the logician pleaded. At this point A said: "Either E is a knave or C is a knight."

Is Oona on this island?

SOLUTIONS

1 · I shall content myself with briefer solutions than those I have formerly given.

Suppose A and B are both knights. Then the common statement they make is true, from which it in turn follows that Oona is on the island. So if they are both knights, then Oona is on the island. This is what they said, so they are both knights. Hence Oona is on the island.

2 · If either one is a knight, then the statement he made is true, from which it in turn follows that Oona is on the island. Therefore, if either one is a knight, Oona is on the island. Thus the statement they both made is true, hence both are knights, so surely at least one is a knight. From this and the truth of the statement they made, it follows that Oona is on the island.

3 · This is an example of what I call a *metapuzzle:* You are not told what B said, but you *are* told that from what A and B said, the logician was able to determine whether Oona was on the island. (If I had not told you that, then you couldn't possibly solve the problem!)

I will first show you that if B had said: "A is a knave, and Oona is not on this island," then the logician couldn't possibly have solved the problem. So suppose that B had said that. Now, A couldn't possibly be a knight, for if he were, then B would be a knight (as A said), which would make A a knave (as B said). Therefore A is definitely a knave. But now it could either be that B is a knight and Oona isn't on the island, or that B is a knave and Oona is on the island, and there is no way to tell which. So if B had said that, the logician couldn't have known whether Oona was on the island. But we are given that the logician did know, hence B *didn't* say that. He must have said: "A is a knave and Oona *is* on this island." Now let's see what happens.

A must be a knave for the same reason as before. If Oona is on the island, we get the following contradiction: It is then true that A is a knave and Oona is on the island, hence B made a true statement, which makes B a knight. But then A made a true statement in claiming that B is a knight and Oona is on the island, contrary to the fact that A is a knave! The only way out of the contradiction is that Oona is *not* on the island. So Oona is not on this island (and, of course, A and B are both knaves).

4 · Since all six natives said the same thing, then they are either all knights or all knaves (all knights, if what they said is true, and all knaves otherwise). Suppose they were all knights. Then it would be true that at least one knave on the island had seen Oona land, but this would be impossible, since none of them are knaves. Therefore they must all be knaves. Hence what they said is false, which means that no knave on the island saw Oona land that evening. But since all the natives are knaves, then no native at all saw Oona land that evening.

5 . We will show that if A is a knave, we get a contradiction: Suppose A is a knave. Then his second statement was false, hence E must be a knight and C must be a knave. Since E is a knight, his statement is true, hence either D is a knave or C is a knight. But C isn't a knight, so D must be a knave. Hence D's statement was false, so either Oona is here today or she was here yesterday. But Oona was not here yesterday (because C said she was and C is a knave), hence she is here today. But this makes A's first statement true, contrary to the assumption that A is a knave! Then A can't be a knave; he must be a knight. Therefore A's first statement was true, so Oona is on this island.

· 5 ·

An Interplanetary Tangle

ON GANYMEDE—a satellite of Jupiter—there is a club known as
the Martian-Venusian Club. All the members are either from Mars
or from Venus, although visitors are sometimes allowed. An earth-
ling is unable to distinguish Martians from Venusians by their ap-
pearance. Also, earthlings cannot distinguish either Martian or
Venusian males from females, since they dress alike. Logicians,
however, have an advantage, since the Venusian women always tell
the truth and the Venusian men always lie. The Martians are the
opposite; the Martian men tell the truth and the Martian women
always lie.

One day Oona and her logician-husband visited Ganymede and
were told about this club. "I'll bet you *I* can tell the men from the
women and the Martians from the Venusians," said the husband
proudly to his wife.

"How?" asked Oona.

"We'll visit the club tonight, which is visitors' night, and I'll
show you!" said the husband (whose name, by the way, was
George).

1

They visited the club that night. "Now, let's see what you can do,"
said Oona somewhat skeptically. "That member over there. Can you

28

tell whether he or she is male or female?" George then went over to the member and asked him or her a single question answerable by yes or no. The member answered, and George then determined whether the member was male or female, though he could not determine whether the member was from Mars or Venus.

What question could it have been?

2

"Very clever!" said Oona, after George had explained the solution. "Now, suppose that instead of wanting to find out whether the member was male or female, you had wanted to find out whether he or she was from Mars or Venus. Could you have done that by asking only one yes-no question?"

"Of course!" said George. "Don't you see how?"

Oona thought about it for a bit and suddenly saw how. How?

3

"If you are *really* clever," said Oona, "you should be able to find out in only one question whether a given member is male or female *and* where the member is from. Let's see you do both things at once using only one yes-no question!"

"Nobody is *that* clever!" replied George. Why did George say that? (This is essentially a repetition of the last puzzle of Chapter 2 of my book *To Mock a Mockingbird.*)

4

Just then a member walked by and made a statement from which George and Oona (who was by now getting the hang of things) could deduce that the member must be a Martian female. What statement could it have been?

5

The next member who walked by made a statement from which George and Oona could deduce that the member must be a Venusian female. What statement could it have been?

6

What statement could be made by either a Martian male, a Martian female, a Venusian male, or a Venusian female?

News soon spread through the club of George and Oona's cleverness in applying logic to determine the sex and/or race of various members. The owner of the club, an American entrepreneur named Fetter, came over to George and Oona's table to congratulate them. "I would like to try you on still other members," said Fetter, "and see what you can do."

7

Just then two members walked by. "Come join us," said Fetter, who introduced them as Ork and Bog. George asked them to tell him something about themselves, and the two made the following statements.

ORK: Bog is from Venus.
BOG: Ork is from Mars.
ORK: Bog is male.
BOG: Ork is female.

From this information, George and Oona could successively identify both the sex and the race of each of them. What is Ork and what is Bog?

8

"You know," said Fetter, after Ork and Bog had left, "that Martians and Venusians often intermarry, and we have several mixed couples in this club. Here is a couple approaching us now. Let's see if you can tell whether or not it is a mixed couple."

I don't remember the couple's first names, so I will simply call them A and B.

"Where are you from?" Oona asked A.

"From Mars," was the reply.

"That's not true!" said B.

Is this a mixed couple or not?

9

"Here comes another couple," said Fetter. "Again I won't tell you whether they are a mixed couple or not. Let's see if you can figure out which one is the husband."

We will call the two A and B. George asked: "Are you both from the same planet?" Here are their replies.

A: We are both from Venus.

B: That is not true!

Which one is the husband?

10

"Here is another couple," said Fetter. "Again, I won't tell you whether it is a mixed couple or not. Let's see what will happen!"

This time I happen to remember their first names—they were Jal and Tork.

"Where are you each from?" asked George.

"My spouse is from Mars," replied Tork.

"We are both from Mars," said Jal.

This enabled George and Oona to classify both of them completely. Which is the husband and which is the wife, and which planet is each of them from?

SOLUTIONS

1 · The simplest question that works is: "Are you Martian?" Suppose you get the answer yes. The speaker is either telling the truth or lying. If the former, then the speaker is really Martian, and being a truth-telling Martian, must be male. If the speaker is lying, then the speaker is really Venusian, hence is a lying Venusian, hence is again male. So in either case, a yes answer indicates that the speaker is male. A similar analysis (which I leave to the reader) shows that a no answer indicates that the speaker is female.

Of course the question "Are you Venusian?" works equally well; a yes answer then indicates that the speaker is female, and a no answer indicates male.

2 · The question, "Are you male?" works. (I leave the verification to the reader.) Alternatively the question, "Are you female?" works as well.

3 · The reason that it is impossible to design a yes-no question that will definitely determine whether a given member is male or female *and* whether the member is Martian or Venusian is that there are four possibilities for the member—a Martian male, a Martian female, a Venusian male, a Venusian female—but there are only two possible responses to the question: yes or no. And so with only two possible responses, one cannot determine which of four possibilities holds.

4 · A simple statement that would work is: "I am a Venusian male." Obviously the statement can't be true, or we would have the contra-

diction of a Venusian male making a true statement. Hence the statement is false, which means that the speaker is *not* a Venusian male. Since the statement is false, the speaker must then be a Martian female.

5 · This is a bit trickier: One statement that works is: "I am either female or Venusian." (Note: Remember that either-or means *at least one and possibly both;* it does not mean *exactly one.*)

If the statement is false, then the speaker is neither female nor Venusian, hence must be a male Martian. But a male Martian does not make false statements, and so we get a contradiction. This proves that the statement must be true, hence the speaker must be either female or Venusian and possibly both. However, if she is female, she must also be Venusian, and if Venusian, also female, because truth-telling females are Venusian and truth-telling Venusians are female. Therefore the speaker must be both Venusian and female.

Incidentally, if the speaker had made the stronger statement, "I am female *and* Venusian," it would be impossible to determine either the sex or the race of the speaker (all that could be inferred is that the speaker is not a Martian male).

6 · One such statement is, "I am either a Martian male or a Venusian female"—or, even more simply, "I always tell the truth." Any liar or truth teller could say that.

7 · Suppose Ork told the truth. Then Bog would be both male and Venusian, hence Bog must have lied. Suppose, on the other hand, that Ork lied. Then Bog is neither male nor Venusian, hence Bog must be a Martian female, so again Bog must have lied. This proves that regardless of whether Ork told the truth or not, Bog definitely lied.

Since Bog lied, then Ork is neither from Mars nor female, hence Ork must be a Venusian male. Therefore Ork also lied, which means that Bog must be a Martian female. And so the solution is that Ork is a Venusian male and Bog is a Martian female (and all four statements were lies).

8 · Since A claimed to be from Mars, then A must be male (as we saw in the solution of Problem 1), and hence B must be female. If A is truthful, then A really is from Mars, B lied, and being a lying female is also from Mars. If A lied, A is really from Venus, B told the truth, and being female is also from Venus. Therefore this is not a mixed couple; they are both from the same planet.

9 · If A's statement is true, then both are from Venus, hence A is from Venus and A must be female. Suppose A's statement is false. Then at least one of them is from Mars. If A is from Mars, A must be female (since A's statement is false). If B is from Mars, B must be male (since B's statement is true), hence again, A must be female. And so A is the wife and B is the husband.

10 · Suppose that Jal told the truth. Then the two really are both from Mars, hence Tork is from Mars and Tork's statement that Jal is from Mars was true. We thus have the impossibility of a couple from the same planet *both* telling the truth. This cannot be, hence Jal must have lied. Therefore at least one of them is from Venus.

If Jal is from Mars, then it must be that Tork is the one from Venus. But then Tork told the truth in claiming that Jal is from Mars so Tork must be female, hence Jal must be male, and we have the impossibility of a male Martian making a false statement. Therefore Jal cannot be from Mars; Jal must be from Venus. Since Jal lied and is from Venus, Jal must be male. Also, since Jal is not from Mars, Tork lied. Hence Tork is a lying female, and thus from Mars.

In summary, Jal is a Venusian male and Tork is a Martian female.

Part III

KNIGHTS, KNAVES, AND PROPOSITIONAL LOGIC

· 6 ·

A Bit of Propositional Logic

THE LIAR–truth teller puzzles of the last three chapters take on an added significance when looked at in terms of the subject known as *propositional logic* (as we will see in the next chapter). In this chapter we will go over a few of the basics—the logical connectives, truth tables, and tautologies. Readers already familiar with these concepts might pass right on to the next chapter (or perhaps just skim this one as a refresher).

THE LOGICAL CONNECTIVES

Propositional logic, like algebra, has its own symbolism, which is relatively easy to learn. In algebra, the letters x, y, z stand for unspecified numbers; in propositional logic, we use the letters p, q, r, s (sometimes with subscripts) to stand for unspecified propositions.

Propositions can be combined by using the so-called *logical connectives*. The principal ones are:

(1) ∼ (not)
(2) & (and)
(3) v (or)

(4) ⊃ (if-then)
(5) ≡ (if-and-only-if)

An explanation of these follows.

(1) Negation. For any proposition p, by ~p we mean the *opposite* or *contrary* of p. We read ~p as "it is not the case that p"; or, more briefly, "not p." The proposition ~p is called the negation of p; it is true if p is false and it is false if p is true. We can summarize these two facts in the following table, which is called the *truth table* for negation. In this table (as in all the tables that follow), we will use the letter "T" to stand for *truth* and "F" to stand for *falsehood*.

p	~p
T	F
F	T

The first line of the truth table says that if p has the value T (i.e., if p is true), then ~p has the value F. The second line says that if p has the value F, then ~p has the value T. We can also write this as follows:

$$\sim T = F$$
$$\sim F = T$$

(2) Conjunction. For any propositions p and q, the proposition that p *and* q are both true is written "p&q" (sometimes "p∧q"). We call p&q the *conjunction* of p and q. It is true if p and q are both true, but false if either one of them is false. We thus have the following four laws of conjunction:

$$T \ \& \ T = T$$
$$T \ \& \ F = F$$
$$F \ \& \ T = F$$
$$F \ \& \ F = F$$

Thus the following is the truth table for conjunction:

p	q	p&q
T	T	T
T	F	F
F	T	F
F	F	F

(3) Disjunction. For any propositions p and q, we let pvq be the proposition that at least one of the propositions p or q is true. We read pvq as "either p or q—and possibly both." (There is another sense of "or," namely, *exactly one,* but this is not the sense in which we will use the word "or." If p and q both happen to be true, the proposition pvq is taken to be true.) The proposition pvq is called the *disjunction* of p and q. Disjunction has the following truth table:

p	q	pvq
T	T	T
T	F	T
F	T	T
F	F	F

We see that pvq is false only in the fourth case—when p and q are both false.

(4) If-Then. For any propositions p and q, we write p⊃q to mean that either p is false or p and q are both true—in other words, if p is true, so is q. We sometimes read p⊃q as "if p, then q," or "p implies q," or "it is not the case that p is true and q is false." We call p⊃q the *conditional* of p and q. For the conditional, we have the following truth table.

p	q	p⊃q
T	T	T
T	F	F
F	T	T
F	F	T

We note that p⊃q is false only in the second line—the case when p is true and q is false. This perhaps needs some explanation: p⊃q is the proposition that it is *not* the case that p is true and q is false. The only way that it can be false is if it *is* the case that p is true and q is false.

(5) If-and-Only-If. Finally, we let p≡q be the proposition that p and q are either both true or both false, or, what is the same thing, that if either one of them is true, so is the other. We read p≡q as "p is true if and only if q is true," or "p and q are equivalent." (We recall that two propositions are called *equivalent* if they are either both true or both false.) The proposition p≡q is sometimes called the *biconditional* of p and q. Here is its truth table.

p	q	p≡q
T	T	T
T	F	F
F	T	F
F	F	T

Parentheses. We need to use parentheses to avoid ambiguity. For example, suppose I write p&qvr. The reader cannot tell whether I mean that p is true and either q or r is true, or whether I mean that either p and q are both true or r is true. If I mean the former, I should write p&(qvr); if I mean the latter, I should write (p&q)vr.

Compound Truth Tables. By the *truth value* of a proposition is meant its truth or falsity—that is T, if p is true, and F, if p is false.

40

Thus the propositions $2 + 2 = 4$ and London is the capital of England, though different propositions, have the same truth value —namely, T.

Consider now two propositions p and q. If we know the truth value of p and the truth value of q, then we can determine the truth values of p&q, pvq, p⊃q, and p≡q—and also the truth value of ~p (as well as the truth value of ~q). It therefore follows that given any combination of p and q (that is, any proposition expressible in terms of p and q, using the logical connectives), we can determine the truth value of this combination once we are given the truth values of p and q. For example, suppose A is the proposition $(p≡(q\&p))⊃(~p⊃q)$. Given the truth values of p and q, we can successively find the values of q&p, p≡(q&p), ~p, (~p⊃q), and finally (p≡(q&p))⊃(~p⊃q). There are four possible distributions of truth values for p and q, and in each of the four cases we can determine the truth value of A. We can do this systematically by constructing the following table:

p	q	q&p	p≡(q&p)	~p	~p⊃q	(p≡(q&p))⊃(~p⊃q)
T	T	T	T	F	T	T
T	F	F	F	F	T	T
F	T	F	T	T	T	T
F	F	F	T	T	F	F

We see then that A is true in the first three cases and false in the fourth.

Let us consider another example: Let $B = (p⊃q)⊃(~q⊃~p)$, and let us make a truth table for B.

p	q	~p	~q	p⊃q	~q⊃~p	(p⊃q)⊃(~q⊃~p)
T	T	F	F	T	T	T
T	F	F	T	F	F	T
F	T	T	F	T	T	T
F	F	T	T	T	T	T

We see that B is true in all four cases, and is hence an example of what is called a *tautology*.

We can also construct a truth table for a combination of three propositional unknowns—p, q, and r—but now there are eight cases to consider (because there are four distributions of T's and F's to p and q, and with each of these four distributions there are two possibilities for r). For example, suppose C is the expression $(p\&(q\supset r))\supset(r\&\sim p)$. It would have the following truth table:

p	q	r	q⊃r	p&(q⊃r)	∼p	r&∼p	(p&(q⊃r))⊃(r&∼p)
T	T	T	T	T	F	F	F
T	T	F	F	F	F	F	F
T	F	T	T	T	F	F	F
T	F	F	T	T	F	F	F
F	T	T	T	F	T	T	F
F	T	F	F	F	T	F	F
F	F	T	T	F	T	T	F
F	F	F	T	F	T	F	F

We will see that C is false in all eight cases; it is the very opposite of a tautology and is an example of what is called a *contradiction*. There are *no* propositions p, q, and r such that $(p\&(q\supset r))\supset(r\&\sim p)$ is true. (We could have seen this without a truth table by using common sense. Suppose p&(q⊃r) is true. Then how could (r&∼p) be true, since ∼p is false?)

If we make a truth table for an expression in four unknowns—say, p, q, r, and s—there are sixteen cases to consider, and so the truth table will have sixteen lines. In general, for any positive whole number n, a truth table for an expression in n unknowns must have 2^n lines (each time we add an unknown, the number of lines doubles).

TAUTOLOGIES

A proposition is called a *tautology* if it can be established purely on the basis of the truth table rules for the logical connectives. For example, suppose that one person says that it will rain tomorrow and a second person says that it won't. We can hardly expect to tell which one is right by using a truth table. We must wait till tomorrow and then *observe* the weather. But suppose a third person says today: "Either it will rain tomorrow or it won't." Now, that's what I would call a *safe* prediction! Without waiting for tomorrow, and making an observation, we know *by pure reason* that he must be right. His assertion is of the form pv∼p (where p is the proposition that it will rain tomorrow), and for *every* proposition p, the proposition pv∼p must be true (as a truth table will easily show).

The more usual definition of tautology involves the notion of a *formula*. By a formula is meant any expression built from the symbols ∼, &, v, ⊃, ≡ and the propositional variables p, q, r, . . . parenthesized correctly. Here are the precise rules for constructing formulas:

(1) Any propositional variable standing alone is a formula.

(2) Given any formulas X and Y already constructed, the expressions (X&Y), (XvY), (X⊃Y), or (X≡Y) are again formulas, and so is the expression ∼X.

It is to be understood that no expression is a formula unless it is constructed according to rules (1) and (2) above.

When displaying a formula standing alone, we can dispense with the outermost parentheses without incurring any ambiguity—for example, when we say "the formula p⊃q," we mean "the formula (p⊃q)."

A formula in itself is neither true nor false, but only becomes true or false when we *interpret* the propositional variables as standing for definite propositions. For example, if I asked: "Is the formula (p&q) true?", you would probably (and rightly) reply: "It depends on what propositions the letters 'p' and 'q' represent." And so a formula such as "p&q" is sometimes true and sometimes false. On the other hand, a formula such as "pv~p" is *always* true (it is true whatever proposition is represented by the letter "p") and is accordingly called a *tautological* formula. Thus a tautological formula is by definition a formula that is *always* true—or what is the same thing, a truth table for the formula will have only T's in the last column. We can then define a *proposition* to be a tautology if it is expressed by some tautological formula under some interpretation of the propositional variables. (For example, the proposition that it is either raining or not raining is expressed by the formula pv~p, if we interpret "p" to be the proposition that it is raining.)

Logical Implication and Equivalence. Given any two propositions X and Y, we say that X *logically* implies Y, or that Y is a *logical consequence* of X if X⊃Y is a tautology. We say that X is *logically equivalent* to Y if X≡Y is a tautology; or, what is the same thing, if X logically implies Y and Y logically implies X.

SOME TAUTOLOGIES

The truth table is a systematic method of verifying tautologies, but many tautologies can be more quickly recognized by using a little common sense. Here are some examples:

(1) ((p⊃q)&(q⊃r))⊃(p⊃r).

This says that if p implies q, and if q implies r, then p implies r. This is surely self-evident (but can, of course, be verified by a truth table). This tautology has a name—it is called a *syllogism*.

(2) (p&(p⊃q))⊃q.

This says that if p is true, and if p implies q, then q is true. This is sometimes paraphrased: "Anything implied by a true proposition is true."

(3) ((p⊃q)&~q)⊃~p.

This says that if p implies a *false* proposition, then p must be false.

(4) ((p⊃q)&(p⊃~q))⊃~p.

This says that if p implies q and also p implies not q, then p must be false.

(5) ((~p⊃q)&(~p⊃~q))⊃p.

This principle is known as *reductio ad absurdum*. To show that p is true, it suffices to show that ~p implies some proposition q as well as its negation ~q.

(6) ((pvq)&~p)⊃q.

This is a familiar principle of logic: If at least one of p or q is true, and if p is false, then it must be q that is true.

(7) ((pvq)&((p⊃r)&(q⊃r)))⊃r.

This is another familiar principle known as *proof by cases*. Suppose pvq is true. Suppose also that p implies r and that q implies r. Then r must be true (regardless of whether it is p or q that is true —or both).

The reader with little experience in propositional logic should benefit from the following exercise.

Exercise 1. State which of the following are tautologies.

(a) (p⊃q)⊃(q⊃p)
(b) (p⊃q)⊃(~p⊃~q)
(c) (p⊃q)⊃(~q⊃~p)

(d) $(p\equiv q)\supset(\sim p\equiv\sim q)$
(e) $\sim(p\supset\sim p)$
(f) $\sim(p\equiv\sim p)$
(g) $\sim(p\&q)\supset(\sim p\&\sim q)$
(h) $\sim(pvq)\supset(\sim pv\sim q)$
(i) $(\sim pv\sim q)\supset\sim(pvq)$
(j) $\sim(p\&q)\equiv(\sim pv\sim q)$
(k) $\sim(pvq)\equiv(\sim p\&\sim q)$
(l) $(q\equiv r)\supset((p\supset q)\equiv(p\supset r))$
(m) $(p\equiv(p\&q))\equiv(q\equiv(pvq))$

Answers. (a) No, (b) No, (c) Yes, (d) Yes, (e) No! (f) Yes, (g) No, (h) Yes, (i) No, (j) Yes, (k) Yes, (l) Yes, (m) Yes (both $p\equiv(p\&q)$ and $(q\equiv(pvq))$ are equivalent to $p\supset q$).

Concerning (e), many beginners think that no proposition p can imply its negation. This is not so! If p happens to be false, then $\sim p$ is true, hence in that case, $p\supset\sim p$ *is* true. However, no proposition can be *equivalent* to its own negation, and so (f) is indeed a tautology.

Discussion. The significance of tautologies is that they are not only true, but *logically certain.* No scientific experiments are necessary to establish their truth—they can be verified on the basis of *pure reason.*

One can alternatively characterize tautologies without appeal to the notion of formulas. Let us define a *state of affairs* as any classification of all propositions into two categories—*true* propositions and *false* propositions—subject to the restriction that the classification must obey the truth table conditions for the logical connectives (for example, we may not classify pvq as true if p and q are both classified as false). A tautology, then, is a proposition which is true in *every possible state of affairs.*

This is related to Leibniz's notion of other possible worlds. Leibniz claimed that of all possible worlds, this one was the best. Frankly,

I have no idea whether he was right or wrong in this, but the interesting thing is that he considered other possible worlds. Out of this, a whole branch of philosophical logic known as possible world semantics has developed in recent years—notably by the philosopher Saul Kripke—which we will discuss in a later chapter. Given any possible world, the set of all propositions that are true for that world, together with the set of all propositions that are false for that world, constitute the state of affairs holding for *that* world. A tautology, then, is true, not only for this world, but for *all* possible worlds. The physical sciences are interested in the state of affairs that holds for the *actual* world, whereas pure mathematics and logic study *all* possible states of affairs.

· 7 ·

Knights, Knaves, and Propositional Logic

KNIGHTS AND KNAVES REVISITED

We can now introduce a simple but basic translation device whereby a host of problems about liars and truth tellers can be reduced to problems in propositional logic. This device will be crucial in several subsequent chapters.

Let us return to the Island of Knights and Knaves. Given a native P, let k be the proposition that P is a knight. Now, suppose that P asserts a proposition X. In general we do not know whether P is a knight or a knave, nor whether X is true or false. But this much we do know: If P is a knight, then X is true, and conversely, if X is true, then P is a knight (because knaves never make true statements). And so we know that P is a knight if and only if X is true; in other words, we know that the proposition $k \equiv X$ is a true proposition. And so we translate "P asserts X" as "$k \equiv X$."

Sometimes we have more than two natives involved—for example, suppose we have two natives P_1 and P_2. We let k_1 be the proposition that P_1 is a knight; we let k_2 be the proposition that P_2 is a knight. If a third native P_3 is involved, we let k_3 be the proposition that P_3 is a knight, and so forth, for all natives involved. We then translate "P_1 asserts X" as "$k_1 \equiv X$"; we translate "P_2 asserts X" as "$k_2 \equiv X$"; and so on.

Let us now look at the first problem in Chapter 3 (page 15). There are two natives P_1 and P_2 (the husband and the wife) involved. We are given that P_1 asserts that P_1 and P_2 are both knaves; we are to determine the types of P_1 and P_2. Now, k_1 is the proposition that P_1 is a knight, hence $\sim k_1$ is equivalent to the proposition that P_1 is a knave (since each inhabitant is a knight or a knave, but not both). Similarly, $\sim k_2$ is the proposition that P_2 is a knave. Hence the proposition that P_1 and P_2 are both knaves is $\sim k_1 \& \sim k_2$. P_1 is asserting the proposition $\sim k_1 \& \sim k_2$. Then, using our translation device, the *reality* of the situation is that $k_1 \equiv (\sim k_1 \& \sim k_2)$ is true. So the problem can be posed in the following purely propositional terms: Given two propositions k_1 and k_2 such that $k_1 \equiv (\sim k_1 \& \sim k_2)$ is true, what are the truth values of k_1 and k_2? If we make a truth table, we can see that the only case in which $k_1 \equiv (\sim k_1 \& \sim k_2)$ comes out true is when k_1 is false and k_2 is true. (We also saw this by common-sense reasoning when we solved the problem in Chapter 3.) The upshot is that the proposition $(k_1 \equiv (\sim k_1 \& \sim k_2)) \supset \sim k_1$ is a tautology, and so is the proposition $(k_1 \equiv (\sim k_1 \& \sim k_2)) \supset k_2$.

The entire mathematical content of this problem is that for any propositions k_1 and k_2, the following proposition is a tautology: $(k_1 \equiv (\sim k_1 \& \sim k_2)) \supset (\sim k_1 \& k_2)$.

The reader might note as well that the converse proposition $(\sim k_1 \& k_2) \supset (k_1 \equiv (\sim k_1 \& \sim k_2))$ is also a tautology, hence the proposition $k_1 \equiv (\sim k_1 \& \sim k_2)$ is logically *equivalent* to the proposition $\sim k_1 \& k_2$.

Now let us look at the second problem in Chapter 3. Here P_1 asserts that either P_1 or P_2 is a knave. We concluded that P_1 is a knight and P_2 is a knave. The mathematical content of this fact is that the proposition $(k_1 \equiv (\sim k_1 v \sim k_2)) \supset (k_1 \& \sim k_2)$ is a tautology.

The reader might note in passing that the converse is also true, and hence that the proposition $(k_1 \equiv (\sim k_1 v \sim k_2)) \equiv (k_1 \& \sim k_2)$ is a tautology.

The translation of Problem 3 is of particular theoretical significance; in fact, let us consider it in the more general form of Theorem I, Chapter 3 (page 18). We have a native P claiming about a certain proposition q that if P is a knight, then q is true (q could be the proposition that P's wife is a knight, or that there is gold on the island, or any proposition whatsoever). We let k be the proposition that P is a knight. Thus P is claiming the proposition k⊃q, and so the reality of the situation is that k≡(k⊃q) is true. From this we are to determine the truth value of k and q. As we have seen, k and q must both be true. Thus the mathematical content of Theorem I, Chapter 3, is that (k≡(k⊃q))⊃(k&q) is a tautology. Of course this fact is not really dependent on the particular nature of the proposition k; for *any* proposition p and q, the proposition (p≡(p⊃q))⊃(p&q) is a tautology. The converse proposition, (p&q)⊃(p≡(p⊃q)), is also a tautology, because if p&q is true, p and q are both true, then p⊃q must be true, hence p≡(p⊃q) must be true. And so the following is a tautology: (p≡(p⊃q))≡(p&q).

Let us now look at Problem 4, or rather Theorem II, in Chapter 3. Here we have P claiming that P is a knight *if and only if* q. Again we let k be the proposition that P is a knight, and so P is claiming the proposition k≡q. Therefore we know that k≡(k≡q) is true. From this we can determine that q must be true, and so the essential mathematical content of Theorem II, Chapter 3, is that the following is a tautology: (k≡(k≡q))⊃q.

This tautology is what I would call the Goodman tautology, since it arises from Nelson Goodman's problem, which is discussed in Chapter 3.

Exercise 1. Let us consider three inhabitants P_1, P_2, and P_3 of the knight-knave island. Suppose P_1 and P_2 make the following statements.

P_1: P_2 and P_3 are both knights.
P_2: P_1 is a knave and P_3 is a knight.

What types are P_1, P_2, and P_3?

Exercise 2. (a) Is the following a tautology?

$$((k_1\equiv(k_2\&k_3))\&(k_2\equiv(\sim k_1\&k_3)))\supset((\sim k_1\&\sim k_2)\&\sim k_3$$

(b) How does this relate to Exercise 1?

Exercise 3. Show that the proposition p_3 is a logical consequence of the following two propositions:

(1) $p_1\equiv\sim p_2$
(2) $p_2\equiv(p_1\equiv\sim p_3)$

Exercise 4. Suppose P_1, P_2, and P_3 are three inhabitants of the knight-knave island, and that P_1 and P_2 make the following statements:

P_1: P_2 is a knave.
P_2: P_1 and P_3 are of different types.

(a) Is P_3 a knight or a knave?
(b) How does this relate to Exercise 3?

THE OONA PROBLEMS

Many of the Oona problems of Chapter 4 can also be solved by truth tables. Consider, for example, the first one (page 23): We have two natives P_1 and P_2; P_1 asserts that if P_1 and P_2 are both knights, then Oona is on the island. P_2 asserts the same thing. We let O be the proposition that Oona is on the island. Using our translation device, we know that the following two propositions are true.

(1) $k_1\equiv((k_1\&k_2)\supset O)$
(2) $k_2\equiv((k_1\&k_2)\supset O)$

We are to determine whether O is true or false. Well, if you make a truth table for the conjunction of (1) and (2)—i.e., for $(k_1 \equiv ((k_1 \& k_2) \supset O)) \& (k_2 \equiv ((k_1 \& k_2) \supset O))$, you will see that the last column contains T's in those and only those places in which O has the value T. Therefore O must be true.

Exercise 5. Suppose the husband goes looking for Oona and meets two natives A and B on an island and they make the following statements:

A: If B is a knight, then Oona is not on this island.
B: If A is a knave, then Oona is not on this island.

Is Oona on this island?

Exercise 6. On another island, two natives A and B make the following statements:

A: If either of us is a knight, then Oona is on this island.
B: If either of us is a knave, then Oona is on this island.

Is Oona on this island?

Exercise 7. On another island, two natives A and B make the following statements.

A: If I am a knight and B is a knave, then Oona is on this island.
B: That is not true!

Is Oona on this island?

MARTIANS AND VENUSIANS
REVISITED

Many of the Mars–Venus–Female–Male puzzles of Chapter 5 can also be solved by truth tables, although the translation device needed is a bit more complex, and this type of problem is not considered further in the present book.

Let us number the club members in some order P_1, P_2, P_3, etc., and for each number i, let V_i be the proposition that P_i is Venusian, and let F_i be the proposition that P_i is female. Then P_i is *Martian* can be written $\sim V_i$, and P_i is *male* can be written $\sim F_i$. Now, P_i tells the truth if and only if P_i is either a Venusian female or a Martian male, which can be symbolized $(V_i \& F_i) v (\sim V_i \& \sim F_i)$, or more simply, $V_i \equiv F_i$. And so now if P_i asserts a proposition X, the reality of the situation is the following proposition: $(V_i \equiv F_i) \equiv X$.

This then is our translation device. Whenever P_i asserts X, we write down: $(V_i \equiv F_i) \equiv X$.

Consider, for example, Problem 7 of Chapter 5 (page 30)—the case of Ork and Bog. Let P_1 be Ork and P_2 be Bog. Then we are given the following four propositions (after we apply the translation device):

(1) $(V_1 \equiv F_1) \equiv V_2$
(2) $(V_2 \equiv F_2) \equiv \sim V_1$
(3) $(V_1 \equiv F_1) \equiv \sim F_2$
(4) $(V_2 \equiv F_2) \equiv F_1$

The truth values of V_1, F_1, V_2, and F_2 can then be found by a truth table. (There are four unknowns V_1, F_1, V_2, and F_2 involved, and so there are sixteen cases to consider!)

Note: Not all liar–truth teller problems can be solved by the translation devices in this chapter. These devices work fine for problems in which we are told what the speakers say and must then deduce certain facts about them. But for the more difficult type of puzzles in which we are to design a question or a statement to do a certain job, more thought is needed. There are other systematic devices that help in many cases, but this is a topic outside the main line of thought in this book.

ANSWERS TO EXERCISES

1 · All three are knaves.

2 · The proposition is a tautology.

3 · We leave this to the reader.

4 · P_3 is a knight.

5 · Oona is not on the island (and both natives are knights).

6 · Oona is on the island (and both natives are knights).

7 · Oona is on the island (and A is a knight; B is a knave).

· 8 ·

Logical Closure and Consistency

INTERDEFINABILITY OF THE LOGICAL CONNECTIVES

We are now familiar with the five basic logical connectives: $\sim$, &, v, $\supset$, and $\equiv$. We could have started with fewer and defined the others in terms of them. This can be done in several ways, some of which will be illustrated in the following puzzles.

1

Suppose an intelligent man from Mars comes down to Earth and wants to learn our logic. He claims to understand the words "not" and "and," but he does not know the meaning of the word "or." How could we explain it to him, using only the notions of *not* and *and*?

Let me rephrase the problem. Given any proposition p, he knows the meaning of $\sim$p, and given any propositions p and q, he knows the meaning of p&q. What the Martian is looking for is a way of writing down an expression or formula in the letters p and q, using *only* the logical connectives $\sim$ and &, such that the expression is logically equivalent to the expression pvq. How can this be done?

2

Now suppose a lady from the planet Venus comes here and tells us that she understands the meanings of ~ and v, but wants an explanation of &. How can we define & in terms of ~ and v? (That is, what expression in terms of p, q, ~, and v is logically equivalent to p&q?)

3

This time a being from Jupiter comes down who understands ~, &, and v, and wants us to define ⊃ in terms of these. How can this be done?

4

Next comes a being from Saturn who strangely enough understands ~ and ⊃, but does not understand either & or v. How can we explain & and v to him?

5

Now comes a being from Uranus who understands only the connective ⊃. It is not then possible to explain to him what & means, nor what ~ means, but it *is* possible to define v in terms of just the one connective ⊃. How can this be done? (The solution is not at all obvious. That it is possible to do it is part of the folklore of mathematical logic, but I have been unable to find out the logician who discovered it.)

6

It is obvious that ≡ can be defined in terms of ~, &, and v. Show two different ways of doing this.

7 · A Special Problem

Suppose a being from outer space understands the meanings of ⊃ and ≡. It will not then be possible to explain to him or her the meaning of ∼, but it will be possible to explain &. How can this be done? (This discovery is mine, as far as I know.)

Two Other Connectives. We now see that the logical connectives ∼, &, v, ⊃, and ≡ can all be defined from just the two connectives ∼ and &, or alternatively from ∼ and v, or alternatively from ∼ and ⊃. Is there just *one* logical connective from which all five connectives can be defined? This problem was solved in 1913 by the logician Henry M. Sheffer. He defined p|q to mean that p and q are not both true. The symbol "|" is known as the "Sheffer stroke"; we can read p|q as "p and q are incompatible" (at least one is false). He showed that ∼, &, v, ⊃, ≡ are all definable from the stroke symbol. Another logical connective that gives all other connectives is ↓; this symbol is called the symbol for *joint denial.* We read p↓q as "p and q are both false," or "neither p nor q is true."

8

How can ∼, &, v, ⊃, ≡ all be defined from Sheffer's stroke? How can they all be defined from joint denial?

The Logical Constant ⊥. A proposition X is called *logically contradictory* or *logically false* if its negation ∼X is a tautology. For example, for any proposition p, the proposition (p&∼p) is logically false. So is p≡∼p.

We shall use the now standard symbol "⊥" as representing any one particular logical falsehood (which one doesn't matter). This can be regarded as fixed for the remainder of this book—and any other books you might read in which this symbol appears. (The symbol "⊥" is pronounced "eet"; it is the symbol "T" written upside down.)

For any proposition p, the proposition ⊥⊃p is a tautology (because ⊥ is logically false, so ⊥⊃p is true, regardless of whether p is true or false). Thus *every* proposition is a logical consequence of ⊥. (It's a good thing that ⊥ itself isn't true, for if it were, everything would be true and the whole world would explode!)

Many modern formulations of propositional logic build their entire theory on just ⊃ and ⊥, because all the other logical connectives can be defined from these two (see Problem 9 below). This is the course we will adopt because it fits in best with the problems of this book. One then defines T as ⊥⊃⊥. We refer to ⊥ as *logical falsehood* and T as *logical truth* (obviously T is a tautology).

9

How does one define all the logical connectives from ⊃ and ⊥?

LOGICAL CLOSURE

A Logically Qualified Machine. To illustrate the important notion of *logical closure,* let us imagine a computing machine programmed to prove various propositions. Whenever the machine proves a proposition, it prints it out (more precisely, it prints out a *sentence* that expresses the proposition). The machine, if left to itself, will run on forever.

We will call the machine *logically qualified* if it satisfies the following two conditions:

(1) Every tautology will be proved by the machine sooner or later.

(2) For any proposition p and q, if the machine ever proves p and proves p⊃q, then it will sooner or later prove q. (We might think of the machine as *inferring* q from the two propositions p and p⊃q. Of course the inference is valid.)

Logically qualified machines have one very important property, of which the following problem illustrates some examples.

58

10

Suppose a machine is logically qualified.

(a) If the machine proves p, will it necessarily prove $\sim\sim$p?

(b) If the machine proves p and proves q, will it necessarily prove the single proposition p&q?

Logical Closure. There is an old rule in logic called *modus ponens*, which is that having proved p and having proved p⊃q, one can then infer q. A set C of propositions is said to be *closed* under modus ponens if for any propositions p and q, if p and p⊃q are both in the set C, so is the proposition q.

We can now define a set C of propositions to be *logically closed* if the following two conditions hold:

Condition 1. C contains all tautologies.

Condition 2. For any propositions p and q, if p and p⊃q are both in C, so is q.

Thus a logically closed set is a set that contains all tautologies and that is closed under modus ponens. (The reason for the term *logically closed* will soon be apparent.)

To say that a machine is logically qualified is tantamount to saying that the set C of all propositions that the machine can prove is a logically closed set. However, logically closed sets also arise in situations in which no machines are involved. For example, we will be considering mathematical systems in which the set of all propositions provable in the system is a logically closed set. Also, a good part of this book will be dealing with logicians whose set of beliefs is a logically closed set.

Logical Consequence. We have defined Y to be a logical consequence of X if the proposition X⊃Y is a tautology. We shall say that Y is a logical consequence of two propositions X_1 and X_2 if it is a consequence of the proposition X_1&X_2—in other words, if $(X_1$&$X_2)$⊃Y is a tautology, or, what is the same thing, if the proposition $X_1 \supset (X_2 \supset Y)$ is a tautology. We define Y to be a logical conse-quence of X_1, X_2, and X_3 if $((X_1 \& X_2) \& X_3) \supset Y$ is a tautology— or, what is the same thing, if $X_1 \supset (X_2 \supset (X_3 \supset Y))$ is a tautology. More generally, for any finite set of propositions $X_1, \ldots, X_n$, we can define Y to be a logical consequence of this set if the proposition $(X_1 \& \ldots \& X_n) \supset Y$ is a tautology.

The importance of logically closed sets is that they enjoy the following property:

Principle L (the Logical Closure Principle). If C is logically closed, then for any n proposition $X_1, \ldots, X_n$ in C, all logical consequences of these n propositions are also in C.

Discussion. Returning to our example of a logically qualified ma-chine, Principle L tells us that if the machine should ever prove a proposition p, then it will sooner or later prove all logical conse-quences of p; if the machine should ever prove two propositions p and q, it will sooner or later prove all logical consequences of p and q—and so forth, for any finite number of propositions.

The same applies to logicians whose set of beliefs is logically closed. If a logician ever believes p, he will sooner or later believe all logical consequences of p; if he should ever believe p and q, he will believe all logical consequences of p and q; and so on.

11

Why is Principle L correct?

12

Another important property of logically closed sets is that if C is logically closed and contains some proposition p and its negation $\sim$p, then every proposition must be in C.

Why is this so?

CONSISTENCY

The last problem brings us to the important notion of *consistency*. A logically closed set C will be called *inconsistent* if it contains ⊥ and *consistent* if it doesn't contain ⊥.

The following is another important property of logically closed sets.

Principle C. Suppose C is logically closed. Then the following three conditions are all equivalent (any one of them implies the other two):
  (1) C is inconsistent (C contains ⊥).
  (2) C contains *all* propositions.
  (3) C contains some proposition p and its negation ∼p.

Note: Given a set S of propositions that is not logically closed, S is said to be inconsistent if ⊥ is a *logical consequence* of some finite subset $X_1, \ldots, X_n$ of propositions in S. (This is, incidentally, equivalent to saying that *every* proposition is a logical consequence of some finite subset of S.) However, we will be dealing almost exclusively with sets that *are* logically closed.

13

Prove Principle C.

SOLUTIONS

1 · To say that at least one of the propositions p and q is true is to say that it is *not* the case that p and q are both false; in other words, it is not the case that ∼p and ∼q are both true. And so pvq is equivalent to the proposition ∼(∼p&∼q). Since the Martian understands ∼ and &, then he will understand ∼(∼p&∼q). And so you

can say to the Martian: "When I say p or q, all I mean is that it is not the case that not p and not q."

2 · To say that p and q are *both* true is equivalent to saying that it is *not* the case that either p or q is false—in other words, it is not the case that either ~p or ~q is true. And so p&q is logically equivalent to the proposition ~(~pv~q).

3 · This can be done in several ways: On the one hand, p⊃q is logically equivalent to ~pv(p&q). It is also equivalent to ~(p&~q) (it is not the case that p is true and q is false), and to ~pvq.

4 · pvq is logically equivalent to ~p⊃q. And so we can get v in terms of ~ and ⊃. Once we have ~ and v, we can get & by the solution to Problem 2 above. More directly, p&q is equivalent to ~(p⊃~q), as the reader can verify.

5 · This is tricky indeed! The proposition pvq happens to be logically equivalent to (p⊃q)⊃q, as the reader can verify by a truth table.

6 · p≡q is obviously equivalent to (p⊃q)&(q⊃p). It is also equivalent to (p&q)v(~p&~q).

7 · We have already solved this in the last chapter, in connection with the third knight-knave problem, page 16; we showed that p&q is logically equivalent to p≡(p⊃q).

8 · Given Sheffer's stroke, we can get the other connectives as follows: First of all, ~p is logically equivalent to p|p. (The proposition p|p is that at least one of the propositions p or p is false, but since the two propositions p and p are the same, this simply says that p is false.) Now that we have ~, we can define pvq to be ~(p|q). (Since p|q is the proposition that at least one of p or q is false, its negation ~(p|q) is equivalent to saying that they are not both false—i.e., that

at least one is true.) Once we have $\sim$ and v, we can get & (by Problem 2), then $\supset$ and $\equiv$ (by Problems 3 and 6). We thus get $\sim$, &, v, $\supset$, and $\equiv$ from Sheffer's stroke.

If we start with joint denial $\downarrow$, instead of Sheffer's stroke, we proceed as follows: We first take $\sim$p to be p$\downarrow$p. Then we take p&q to be $\sim$(p$\downarrow$q). Having gotten $\sim$ and &, we can then get all the rest in the manner we have seen.

9 · The proposition $\sim$p is logically equivalent to p$\supset\perp$, and so we can get $\sim$ from $\supset$ and $\perp$. Once we have $\sim$ and $\supset$, we can get v and & (by Problem 4). Then we can get $\equiv$ from $\supset$ and &.

10 · (a) Suppose the machine proves p. It will also prove p$\supset\sim\sim$p (since this is a tautology), hence it will print $\sim\sim$p (by the second condition defining logical qualification).

(b) Suppose the machine proves p and proves q. Now, the proposition p$\supset$(q$\supset$(p&q)) is a tautology, hence the machine will prove it. Once the machine has proved p and p$\supset$(q$\supset$(p&q)), it must prove q$\supset$(p&q). Once the machine has proved this, then, since it proves q, it must prove p&q.

(Actually this problem is but a special case of the next, as the reader will see.)

11 · Let us first consider the case n $=$ 1: Suppose X_1 is in C, and Y is a logical consequence of X_1. Then $X_1 \supset$Y is a tautology, hence is in C (by Condition 1). Since X_1 and $X_1 \supset$Y are both in C, so is Y (by Condition 2).

Now let us consider the case n $=$ 2: Suppose X_1 and X_2 are both in C, and Y is a logical consequence of X_1 and X_2. Then $(X_1$&$X_2)\supset$Y is a tautology, hence $X_1 \supset (X_2 \supset$Y) is a tautology (as the reader can verify) and is therefore in C. Since X_1 and $X_1 \supset (X_2 \supset$Y) are both in C, so is $X_2 \supset$Y (Condition 2). Since X_2 and $X_2 \supset$Y are in C, so is Y (again by Condition 2).

For the case n $=$ 3, suppose Y is a logical consequence of X_1, X_2, and X_3. Then $X_1 \supset (X_2 \supset (X_3 \supset Y))$ is a tautology—it is logically equivalent to $(X_1 \& X_2 \& X_3) \supset Y$. Then using Condition 2 three times, we successively get $X_2 \supset (X_3 \supset Y)$ in C, $(X_3 \supset Y)$ in C, and finally Y in C.

It should be obvious how this proof generalizes for any positive whole number n.

Note: We have remarked that Problem 10 is but a special case of the present problem. The reason, of course, is that $\sim\sim p$ is a logical consequence of p, hence any logically closed set containing p must also contain $\sim\sim p$. Also, p&q is a logical consequence of the two propositions p and q, so any logically closed set containing both p and q must also contain p&q.

12 · Suppose that p and its negation $\sim p$ are both in C and that C is logically closed. Let q be any proposition whatsoever. The proposition $(p \& \sim p) \supset q$ is a tautology (as the reader can check by a truth table, or more simply by observing that since $p \& \sim p$ must be false, then $(p \& \sim p) \supset q$ must be true). And so q is a logical consequence of the true propositions p and $\sim p$. Then according to Principle L, the proposition q must also be in C. And so *every* proposition q is in C.

13 · We will show that the three conditions are all equivalent by showing that (1) implies (2), which in turn implies (3), which in turn implies (1).

Suppose (1). Since every proposition is a logical consequence of $\perp$ and $\perp$ is in C, then every proposition is in C (by Principle L). Thus (2) holds.

It is completely obvious that (2) implies (3), because if *all* propositions are in C, then for *any* proposition p, both p and $\sim p$ are in C.

Now suppose that (3) holds—i.e., that for some p, both p and $\sim p$ are in C. Since $\perp$ is a logical consequence of p and $\sim p$, then $\perp$ must be in C (by Principle L)—i.e., C must be inconsistent.

· Part IV ·

LET'S BE
CAREFUL!

· 9 ·

Paradoxical?

WE NOW have the background to embark on our journey to Gödel's consistency predicament, which we will reach in Chapter 12, encountering many interesting problems along the way. Let us start with a problem closely related to one of the variants of the Surprise Examination Paradox in Chapter 2.

We are back on the Island of Knights and Knaves, where the following three propositions hold: (1) knights make only true statements; (2) knaves make only false ones; (3) every inhabitant is either a knight or a knave. These three propositions will be collectively referred to as the "rules of the island."

We recall that no inhabitant can claim that he is not a knight, since no knight would make the false statement that he isn't a knight and no knave would make the true statement that he isn't a knight.

Now suppose a logician visits the island and meets a native who makes the following statement to him: *"You will never know that I am a knight."*

Do we get a paradox? Let us see. The logician starts reasoning as follows: "Suppose he is a knave. Then his statement is false, which means that at some time I *will* know that he is a knight, but I can't *know* that he is a knight unless he really is one. So, if he is a knave, it follows that he must be a knight, which is a contradiction. Therefore he can't be a knave; he must be a knight."

So far, so good—there is as yet no contradiction. But then he continues reasoning: "Now I know that he is a knight, although he

67

said that I never would. Hence his statement was false, which means that he must be a knave. Paradox!"

Question. Is this a genuine paradox?

Discussion. This problem bears a good deal of analysis! To begin with, the paradox (if it really is one) is what would be called a *pragmatic* paradox rather than a purely logical one, since it involves not only logical notions such as truth and falsity, but pragmatic notions such as *knowing.* To emphasize the pragmatic nature of the question, is it not possible that whether or not a paradox arises might depend on the person to whom the statement is made? It certainly is! As an extreme example, the native could surely say it to a dead person, and no paradox would arise. (He points to the corpse and says: "You will never know that I'm a knight." Well, he is certainly right; the corpse will indeed never know that he is a knight, and so the native is in fact a knight—since what he said is true. But since the corpse will never know it, no contradiction arises.) To take a less extreme example, the native might say this to a person who is alive but deaf and hence does not hear the statement; again, no paradox would arise.

So we must assume that the visitor to the island was alive and heard the statement, but this is still not enough. We must assume a certain *reasoning* ability on the visitor's part, because if the visitor has no reasoning ability, he would not go through the argument I have given. (The native would say: "You will never know that I'm a knight." The visitor might then say: "That's interesting," walk away, and never think about the matter again. Hence no paradox would arise.) And so we must make explicit what reasoning abilities the logician has.

We will define an individual to be a *reasoner of type 1* if he thoroughly understands propositional logic—that is, if the following two conditions hold:

(1) He believes all tautologies.

(2) For any propositions X and Y, if he believes X and believes X⊃Y, then he believes Y.

In the terminology of Chapter 8, the set of beliefs of a reasoner of type 1 is logically closed. It then follows by Principle L of Chapter 8, that given any finite set S of propositions that he believes, he must believe all logical consequences of S as well.

We shall now make the assumption that the visitor to the island is a reasoner of type 1. Of course this assumption is highly idealized, since there are *infinitely* many tautologies, hence our assumption implies something like immortality on the reasoner's part. However, little things like that don't bother us in the timeless realm of mathematics. We simply imagine the reasoner so programmed that (1) sooner or later he will believe every tautology; (2) if he ever believes p and ever believes p⊃q, then sooner or later he will believe q. It then follows by Principle L of Chapter 8 that given any finite set S of propositions, if the reasoner believes all the propositions in S, then for any proposition Y that is a logical consequence of S, the reasoner will sooner or later believe Y.

We still need further assumptions. For one thing we must assume a certain *self-consciousness* on the reasoner's part; specifically, we must assume that if the reasoner ever knows something, then he knows that he knows it (otherwise he could never have said, "Now I know that he's a knight," and my argument wouldn't go through).

Considering the problem with all these assumptions, does a genuine paradox now arise? Still not, for although I told you that the rules of the island hold (every inhabitant is either a knight or a knave; knights make only true statements; knaves make only false statements), we must make the additional assumption that the reasoner *believes* that the rules of the island hold. Indeed, it is perfectly reasonable that the reasoner might believe this at the outset, but after finding himself in a contradiction following the argument I gave, he would have rational grounds for *doubting* the rules of the island. (I imagine you and I would do just that after finding ourselves in such a predicament!) Well, to make the problem really interesting, let us make as our final assumption that the reasoner believes *and continues to believe* the rules of the island.

1.

Now we run into an interesting problem. Under the additional assumptions we have made, the reasoner's argument that the native is a knight and that the native is a knave seems perfectly valid. Yet the native can't be both a knight and a knave! So what is wrong with the reasoner's argument?

Solution. I phrased the above problem in a very misleading way. (Occasionally I feel like being a bit sneaky!) It is not the reasoner's argument that was wrong; it's that the situation I described could never have arisen. If a reasoner of type 1 comes to a knight-knave island and believes the rules of the island (and hears whatever statements are made to him), then it is logically impossible that any native will say to him, "You will never know that I am a knight."

To prove this, we don't need one of the assumptions we have made—namely, that if the reasoner knows something, then he knows that he knows it. Even without this assumption, we can get a contradiction as follows: The reasoner reasons, "Suppose he is a knave. Then his statement is false, which means that I will know that he is a knight, which implies that he really is a knight. Therefore the assumption that he is a knave leads to a contradiction, so he must be a knight."

Without going any further in the logician's reasoning process, we can derive a contradiction. The logician has so far reasoned correctly and has come to the conclusion that the native is a knight. Since he has reasoned correctly, then the native really is a knight, and so the reasoner *knows* that the native is a knight. However, the native said he would never know that, hence the native must be a knave.

Therefore the native is both a knight and a knave, which is a contradiction.

Instead of using the word "know," we could just as well have said "correctly believe," and our argument would still go through. We say that an individual *correctly believes* a proposition p if he believes p and p is true.

We have now proved Theorem I.

Theorem I. Given a knight-knave island and a reasoner of type 1 who believes the rules of the island (and who hears any statement addressed to him), it is logically impossible that any native can say to him, "You will never correctly believe that I am a knight."

Discussion. Some critical readers might object to the proof I have given of the above theorem on the grounds that I have credited the reasoner with more abilities than I have explicitly ascribed to him —namely, that the reasoner makes assumptions and subsequently discharges them. In the specific case in hand, the reasoner started out: "Suppose he is a knave. Then —— ." Well, this is really only a matter of convenience, not necessity. I could have given the reasoner's argument in the following, more direct form: "If he is a knave, then his statement is false. If his statement is false, then I'll correctly believe he is a knight. If I correctly believe he is a knight, then he is a knight. Putting these three facts together, if he is a knave, then he is a knight. From this last fact it logically follows that he is a knight."

It is a common practice in logic to prove a proposition of the form p⊃q (if p then q) by supposing that p is true and then trying to derive q. If this can be done, then the proposition p⊃q is established. In other words, if assuming p as a premise leads to q as a conclusion, then the proposition p⊃q has been proved. (This technique is part of what is known as *natural deduction*.) The whole point is that anything that can be proved using this device can also be proved without it. (There is a well-known theorem in logic to this effect; it is called the *deduction theorem.*) And so we shall allow our "reasoners" to use natural deduction; a reasoner of type 1 who does this cannot prove any more facts than he can without using natural deduction, but the proofs using natural deduction are usually shorter and easier to follow. Therefore we shall continue to let our reasoners use natural deduction.

2 · A Dual Problem

We continue to assume that the reasoner is of type 1, that he believes the rules of the island, and that he hears all remarks addressed to him.

Suppose the native, instead of saying, "You will never correctly believe I'm a knight," says, "You will correctly believe I'm a knave."

Do we then get a contradiction? (The reader should try solving this before reading the solution.)

Solution. The reasoner reasons as follows: "Suppose he is a knight. Then his statement is true, which means that I will correctly believe he is a knave, which in turn implies that he is a knave. Hence the assumption that he is a knight leads to a contradiction, therefore he must be a knave."

At this point the reasoner believes the native is a knave, and he has reasoned correctly, hence the native is a knave. On the other hand, since the reasoner correctly believes that the native is a knave, the native's statement was true, which makes him a knight. So we do indeed get a contradiction.

SOME RELATED PROBLEMS

Let us now leave the Island of Knights and Knaves for a while and consider a problem related to the paradox of Chapter 3. A student asks his theology professor: "Does God really exist?" The professor gives the following curious answer: "God exists if and only if you don't correctly believe that He does."

3

Suppose that the student is a reasoner of type 1 and that the professor's statement is true and that the student believes the statement. Do we then get a paradox?

Solution. Yes, we do! To begin with, even forgetting that the student is a reasoner of type 1 and that he believes the professor's statement, it follows that God must exist, because if God didn't exist, then the student *would* correctly believe that God exists, but no one can *correctly* believe a false proposition. Therefore God must really exist (assuming that the professor's statement is true).

Now, the student, being a reasoner of type 1, knows propositional logic as well as you or I, hence he also is able to reason that if the professor's statement is true, then God must exist. But he also believes the professor's statement; therefore he must believe that God exists. And since we have proved that God exists (under the three assumptions of the problem), then the student *correctly* believes that God exists. But God exists if and only if the student *doesn't* correctly believe that God exists. From this it follows that God doesn't exist if and only if the student *does* correctly believe that God exists. (For any proposition p and q, the proposition $p \equiv \sim q$ is logically equivalent to the proposition $\sim p \equiv q$.) Since God doesn't exist if and only if the student correctly believes that God exists, and the student does correctly believe that God exists, then it follows that God doesn't exist. Thus the three assumptions of the problem lead to the paradox that God does exist and doesn't exist.

Of course the same paradox would arise if the professor had instead said: "God exists if and only if you correctly believe that He doesn't exist." We leave the proof of this as an exercise for the reader.

It is now important for us to realize that the above paradox is essentially the same as that of Problem 1 concerning the knight-knave island, although they may appear different. The seeming differences are: (1) In the above paradox, the professor made an "if and only if" statement, whereas in Problem 1, the native did not; he said outright that the reasoner would never correctly believe that the native is a knight. (2) In the above paradox, the student believes the professor, whereas in Problem 1, the reasoner has no initial belief that the native is a knight. However, these two differences in a sense cancel each other out, as we will now see. The key to this is the translation device in Chapter 7.

In virtually all the problems that follow, we will be dealing with only two individuals—the native of the island who makes the statement, and the reasoner who hears the statement. We will consistently use the letter k for the proposition that the native in question is a knight. Then, as we saw in Chapter 7, whenever the native asserts a proposition q, the proposition $k \equiv q$ is true. Now, the reasoner *believes* the rules of the island (he believes that knights make true statements and knaves make false ones), and we are assuming that he hears any statement made to him. Therefore, whenever the native asserts a proposition q to the reasoner, the reasoner *believes* the proposition $k \equiv q$. Indeed, from now on, when we say that the rules of the island hold, we need mean no more than that for any proposition q, if the native asserts q, then the proposition $k \equiv q$ is true. And when we say that the reasoner believes the rules of the island (and hears all statements made to him), we need mean no more than that for any proposition q, if the native asserts q to the reasoner, then the reasoner believes the proposition $k \equiv q$.

For any proposition p, we let Bp be the proposition that the reasoner believes (or will believe) p. And we let Cp be the proposition p&Bp. We read Cp as "The reasoner *correctly* believes p."

Now, in Problem 1, the native asserted the proposition $\sim Ck$ ("You will never correctly believe that I am a knight"). Since the rules of the island hold, then the proposition $k \equiv \sim Ck$ is true. Since the reasoner believes the rules of the island, then he believes the proposition k≡~Ck. These two facts turn out to be logically incompatible, hence a paradox arises.

In Problem 3, let g be the proposition that God exists. The professor asserted outright the proposition g≡~Cg. Under the assumption that the professor makes only true statements, the proposition g≡~Cg must be true. Since the student believed the professor, then he believed the proposition g≡~Cg. Again, the truth of g≡~Cg turned out to be logically incompatible with the student believing g≡~Cg (since the student is a reasoner of type 1).

We now see exactly what the two paradoxes have in common; in both cases we have a proposition p (which is k, for Problem 1, and g for Problem 3) such that p≡~Cp is both true and believed by the reasoner—in other words, it is *correctly* believed by the reasoner, and this is logically impossible if the reasoner is of type 1. Thus both paradoxes (or rather their resolutions that the given conditions are logically incompatible) are special cases of the following theorem.

Theorem A. There is no proposition p such that a reasoner of type 1 can correctly believe the proposition p≡~Cp. In other words, there is no proposition such that a reasoner of type 1 can correctly believe: "The proposition is true if and only if I don't correctly believe that it is true."

The proof of Theorem A is little more than a repetition of the two special cases already considered, but it may help to consider it in a more general setting and to point out some of its interesting features.

To begin with, for any propositions p and q, the proposition (p≡~(p&q))⊃p is a tautology (as the reader can verify). In particular, the proposition (p≡~(p&Bp))⊃p is a tautology. We are letting Cp be the proposition p&Bp, and so (p≡~Cp)⊃p is a tautology. Suppose now that a reasoner of type 1 correctly believes p≡~Cp, we then get the following contradiction: Since the reasoner *correctly* believes p≡~Cp, then p≡~Cp must be true. Also (p≡~Cp)⊃p is true (it is a tautology), and so p must be true. Now since the reasoner is of type 1, he *believes* the tautology $(p\equiv\sim Cp)\supset p$ and he also believes $p\equiv\sim Cp$ (by assumption), and since he is of type 1, he will then believe p. And so p is true, and he believes p, so he correctly believes p. Thus Cp is true, hence $\sim Cp$ is false. But since p is true and $\sim Cp$ is false, it cannot be that $p\equiv\sim Cp$ (since a true proposition cannot be equivalent to a false proposition), and so we get a contradiction from the assumption that a reasoner of type 1 correctly believes $p\equiv\sim Cp$.

There is also the following "dual" of Theorem A whose proof we leave to the reader.

Theorem A°. There is no proposition p such that a reasoner of type 1 can correctly believe $p\equiv C(\sim p)$.

Exercise 1. Prove Theorem A°.

Exercise 2. Suppose we have a perfectly arbitrary operation B which assigns to every proposition p a certain proposition Bp. (What the proposition Bp is needn't be specified; in this chapter, we have let Bp be the proposition that the reasoner *believes* p; in a later chapter, in which we will be discussing mathematical systems rather than reasoners, Bp will be the proposition that p is *provable* in the system. But for now, Bp will be unspecified.) We let Cp be the proposition (p&Bp).

(a) Show that one can derive a logical contradiction from the following assumptions:

(i) All propositions of the form BX, where X is a tautology.

(ii) All propositions of the form $(BX\&B(X\supset Y))\supset BY$.

(iii) Some proposition of the form $C(p\equiv\sim Cp)$.

(b) Show that we also get a logical contradiction if we replace (iii) with "Some proposition of the form $C(p\equiv C\sim p)$."

(c) Why is Theorem A a special case of (a) above? Why is Theorem A° a special case of (b)?

76

· 10 ·

The Problem Deepens

CONCEITED REASONERS

We are back to the Island of Knights and Knaves. Suppose, now, that the native, instead of saying: "You will never correctly believe I'm a knight," makes the following statement: *"You will never believe that I am a knight."*

The native has left out the word "correctly," and as a result things will get far more interesting. We continue to assume that the one addressed is a reasoner of type 1 and that he believes the rules of the island (and also that he has heard the statement) and that the rules of the island really hold. And now we shall make the further assumption that the reasoner is completely accurate in his judgments; he doesn't believe any proposition that is false. Do we still get a paradox?

Well, suppose the native is a knave. Then his statement is false, which means that the reasoner will believe he is a knight. And since the reasoner is accurate in his judgments, then the native really is a knight. Thus the assumption that the native is a knave leads to a contradiction, so the native must be a knight.

Now, the reasoner is of type 1 and knows as much logic as you and I. What is to prevent him from going through the same reasoning process that we just went through and coming to the same conclusion—namely, that the native must be a knight? Therefore the reasoner will *believe* that the native is a knight, which makes the native's statement false, hence the native must be a knave. But we have already proved that the native is a knight. Paradox!

1

The above argument is fallacious! Can the reader spot the fallacy? (Hint: Despite the fact that the reasoner knows propositional logic as well as you and I, there is something we know that the reasoner doesn't know. What is it?)

Solution. I told you that the reasoner is always accurate; I never said that he *knew* he was accurate! If he knew he was accurate (which in fact he can't know), we would get a paradox. You see, part of our proof that the native is a knight used the assumption that the reasoner is always accurate; if the reasoner made the same assumption, then he could likewise prove that the native is a knight, thus making the native a knave.

Let us now retract the assumption that the reasoner is always accurate in his judgments, but let us suppose that the reasoner *believes* that he is always accurate. Thus for any proposition p, the reasoner believes that if he should ever believe p, then p must be true. Such a reasoner we will call a *conceited* reasoner. Thus a conceited reasoner is one who believes that he is incapable of believing any false proposition.

And so we retract the assumption that the reasoner is always accurate and replace it with the assumption that the reasoner *believes* that he is always accurate. Do we then get a paradox? No, we don't; we instead have the following more interesting result.

Theorem 1. Suppose a native of a knight-knave island says to a reasoner of type 1: "You will never believe I'm a knight." Then if the reasoner believes himself always accurate, he will lapse into an inaccuracy—i.e., he will sooner or later believe something false.

2

Prove Theorem 1.

Solution. The reasoner reasons: "Suppose he is a knave. Then his statement is false, which means that I will believe he is a knight. But if I ever believe he is a knight, he must really be one, because I am not capable of making mistakes [sic!]. So if he is a knave, he is a knight, which is not possible. Therefore he is not a knave; he is a knight."

At this point, the reasoner believes the native is a knight. Since the native said that the reasoner would never believe that, then the native is in fact a knave. So the reasoner now has the *false* belief that the native is a knight.

The interesting thing is that if the reasoner had been more modest and had not assumed his own infallibility, he would never have been driven into the inaccuracy of believing the native a knight. The reasoner has been justly punished for his conceit!

PECULIAR REASONERS

Let us say that a reasoner is *accurate* with respect to a given proposition p if the reasoner's believing p implies that p is true; in other words, if it is either not the case that he believes p, or it is the case that he believes p and p is true. We will say that the reasoner is inaccurate with respect to p if he believes p and p is false.

It is a noteworthy fact about the last problem that the reasoner was inaccurate with respect to the very proposition about which he believed himself to be accurate—namely, the proposition that the native is a knight. By believing that he was accurate with respect to that proposition, he finally came to believe that the native was a knight, thus making the native a knave. Of course our proof that the native is a knave rested on our assumption that the rules of the island held. Suppose we retract this assumption but continue to assume that the reasoner *believes* that the rules of the island hold; does it still follow that the reasoner will believe some false proposition? Of course the reasoner will still believe that the native is a knight, although the native said he never would; but if the rules of the island don't hold, then the native is not necessarily a knave. However, even though our proof of Theorem 1 fails if we retract the assumption that the rules of the island hold, we have the following more startling proposition:

Theorem 2. Suppose an inhabitant of the island says to a reasoner of type 1: "You will never believe that I'm a knight." Suppose the reasoner *believes* that the rules of the island hold. Then regardless of whether the rules really hold or not, if the reasoner is conceited, he will come to believe some false proposition.

3

Under the assumption of Theorem 2, what false proposition will the reasoner believe?

Solution. By the same method of proof we followed for Theorem 1, the reasoner will come to believe that the native is a knight. This belief is not necessarily false (since the rules of the island don't necessarily hold), but then the reasoner continues: "Since he is a knight and he said I would never believe he is a knight, then what he said must be true—namely, that I don't believe (i.e., it's not the case that I believe) he's a knight. So I don't believe he is a knight."

At this point, the reasoner believes that the native is a knight and also believes that he doesn't believe that the native is a knight. So he has the false belief that he doesn't believe the native is a knight (it's false, since he *does* believe that the native is a knight!).

The conclusion of the above problem is really quite weird. The reasoner believes that the native is a knight and also believes that he doesn't believe that the native is a knight. Now, this does not involve a logical inconsistency on the part of the reasoner, though it certainly does involve a psychological peculiarity. We shall call a reasoner *peculiar* if there is some proposition p such that he believes p and also believes that he doesn't believe p. This condition of course implies that the reasoner is inaccurate (because he believes the false proposition that he doesn't believe p). And so a peculiar reasoner is automatically inaccurate, but not necessarily inconsistent.

Let us say that a reasoner is *peculiar with respect to a given proposition p* if he believes p and also believes that he doesn't believe p. A reasoner is then peculiar if and only if there is at least one proposition p with respect to which he is peculiar. If a reasoner is peculiar with respect to p, then he is inaccurate, not necessarily with respect to p, but with respect to Bp.

We now see that even if we remove the assumption that the rules of the island actually hold, if the reasoner believes that they hold and he is of type 1 and a native tells him that he will never believe the native is a knight, then if the reasoner believes that he is accurate with respect to the proposition that the native is a knight, his belief forces him to be inaccurate with respect to the proposition that he *believes* the native is a knight. If the rules of the island do actually hold, then the reasoner will also be inaccurate with respect to the proposition that the native is a knight— i.e., he will believe that the native is a knight, whereas the native is really a knave.

Of course this same problem can be formulated in the context of the student and his theology professor. Suppose the student is a reasoner of type 1 and his professor says to him: "God exists if and only if you will never believe that He does." Suppose also that the student believes that if he ever believes that God exists, then God does exist. This will mean: (1) If the student believes the professor, then he will wind up believing that God exists and also believing that he doesn't believe that God exists. (2) If the professor's statement is also true, then God doesn't exist, but the student will believe that God does exist.

Both versions of this problem are special cases of the following theorem.

Theorem A. Suppose a reasoner is of type 1 and that there is a proposition p such that he believes the proposition $p \equiv \sim Bp$ and also believes that $Bp \supset p$. Then it follows that:

(a) He will believe p and also believe that he doesn't believe p (he will be peculiar with respect to p).

(b) If also $p \equiv \sim Bp$ is true, then p is false, but he will believe p.

Proof. (a) He believes $p \equiv \sim Bp$, hence he believes $Bp \supset \sim p$ (which is a logical consequence of $p \equiv \sim Bp$). He also believes $Bp \supset p$ (by hypothesis), hence he must believe $\sim Bp$ (which is a logical consequence of $Bp \supset \sim p$ and $Bp \supset p$). But he also believes $p \equiv \sim Bp$, hence he must believe p (which is a logical consequence of $\sim Bp$ and $p \equiv \sim Bp$). And so he believes p and he also believes $\sim Bp$ (he believes that he doesn't believe p!).

(b) Suppose also that $p \equiv \sim Bp$ is true. Since $\sim Bp$ is false (he *does* believe p!), then p, being equivalent to the false proposition $\sim Bp$, is also false. Therefore he believes p, but p is false.

Exercise. Suppose a native says to a conceited reasoner of type 1: "You will believe that I am a knave." Prove: (a) If the reasoner believes that the rules of the island hold, then he will believe that the native is a knave and also that he doesn't believe that the native is a knave. (b) If also the rules of the island really hold, then the native is in fact a knight.

REASONERS OF TYPE 1*

By a reasoner of *type 1**, we shall mean a reasoner of type 1 with the added property that for any propositions p and q, if he ever believes the proposition p⊃q, then he will believe that if he ever believes p then he will also believe q. In symbols, if he ever believes p⊃q, then he will also believe Bp⊃Bq.

Let us note that if a reasoner of type 1 does believe p⊃q, then it is true that if he ever believes p, then he will believe q—i.e., Bp⊃Bq is a true proposition (if he believes p⊃q). What a reasoner of type 1* has that a reasoner of just type 1 doesn't have is that if he believes p⊃q, then not only is the proposition Bp⊃Bq a true one, but he correctly *believes* Bp⊃Bq. Thus a reasoner of type 1* has a shade more "self-awareness" than a reasoner who is only type 1.

We continue to assume that the reasoner, who is now of type 1*, believes the rules of the island and hears all statements made to him, and so whenever the native asserts a proposition p, the reasoner believes the proposition k≡p, where k is the proposition that the native is a knight.

The following fact will be quite crucial:

Lemma 1.[†] Suppose the native asserts a statement to a reasoner of type 1*. Then the reasoner will believe that if he ever believes that the native is a knight, he will also believe what the native said.

Problem 4. How is this lemma proved?

Solution. (The proof is really quite simple!) Suppose the native asserts the proposition p to the reasoner. Then the reasoner believes

---

[†]A lemma is a proposition proved not so much for its own sake as for help in proving subsequent theorems. You might say that a lemma is a proposition that is not "dignified enough" to be called a theorem.

the proposition $k \equiv p$. Then he also believes $k \supset p$, because $k \supset p$ is a logical consequence of $k \equiv p$. Then, since he is of type 1*, he will believe $Bk \supset Bp$.

I have called a reasoner "conceited" if he believes in his own infallibility. I would hardly regard a person's belief that he is not peculiar as an act of conceit; indeed, to assume that one is not peculiar is a perfectly reasonable assumption. I'm not even sure whether it is psychologically possible for a person to be peculiar. Could a person really believe something and also believe that he doesn't believe it? I doubt it. Yet it is not *logically* impossible for a person to be peculiar.

At any rate, to have confidence in one's own nonpeculiarity is far more reasonable than to have confidence in one's complete accuracy. Therefore the following theorem is a bit sad.

Theorem 3. Suppose a native says to a reasoner of type 1*: "You will never believe that I am a knight." Then if the reasoner believes that he is not (and never will be) peculiar, he will become peculiar!

Problem 5. Prove Theorem 3.

Solution. The proof of this is a bit more elaborate than any other proof so far.

Assume that the native makes this statement and that the reasoner believes that he is incapable of being peculiar. Since the native made the statement, then by Lemma 1, the reasoner will believe that if he ever believes that the native is a knight, he will also believe what the native said. And so the reasoner reasons: "Suppose I should ever believe that he is a knight. Then I'll believe what he says—i.e., I'll believe that I don't believe he is a knight. And so I will then believe he's a knight and I will also believe that I don't believe he is a knight. This means I will be peculiar. Therefore, if I ever believe he's a knight, I will become peculiar. Since I will never be peculiar [sic], then I will never believe that he's a knight. Since he said I wouldn't, his statement is true, and so he is a knight."

At this point the reasoner has come to the conclusion that the native is a knight, and a bit earlier he came to the conclusion that he doesn't believe that the native is a knight. He has thus lapsed into peculiarity.

Of course the above proposition can be stated and proved in the following more general form:

Theorem B. For any reasoner of type 1*, if he believes any proposition of the form $p \equiv \sim Bp$ ("p is true if and only if I will never believe p"), then he cannot believe that he is not peculiar, unless he lapses into peculiarity.

Moral. If you are a reasoner of type 1*, and you wish to believe that you are not peculiar, you can avoid becoming peculiar by simply refusing to believe any proposition of the form "p if and only if I will never believe p."

In particular, if you ever visit the knight-knave island (or what you have been told is a knight-knave island) and a native tells you that you will never believe that he is a knight, then your wisest course is to refuse to believe that the rules of the island hold.

Later in this book, however, when we come to the study of mathematical systems and talk about provability in the system rather than beliefs of a reasoner, we will see that the analogue of the option of not believing $p \equiv \sim Bp$ will not be open.

Part V

THE CONSISTENCY PREDICAMENT

· 11 ·

Logicians
Who Reason About
Themselves

We are getting close to Gödel's consistency predicament. But first, we need to consider reasoners of higher degrees of self-awareness than those of just type 1.

ADVANCING STAGES OF
SELF-AWARENESS

We will now define reasoners of types 2, 3, and 4, which represent advancing degrees of self-awareness. Reasoners of type 4 play a major role in the dramas that will unfold.

Reasoners of Type 2. Suppose a reasoner of type 1 believes p and believes p⊃q. Then he will believe q. This means that the proposition (Bp&B(p⊃q))⊃Bq is *true* for a reasoner of type 1. However, the reasoner doesn't necessarily *know* that this proposition is true. Well, we define a reasoner to be of type 2 if he is of type 1 and *believes* all propositions of the form (Bp&B(p⊃q))⊃Bq. (He "knows" that his set of beliefs—past, present, and future—is closed under modus ponens. For any propositions p and q, he believes: "If I should ever believe p and believe p⊃q, then I will also believe q.")

To emphasize a point, reasoners of type 2 have a certain "self-awareness" not necessarily present in reasoners of type 1. A reasoner of type 1 who believes p and believes p⊃q will sooner or later believe q; reasoners of type 2 also *know* that if they ever believe p and believe p⊃q, they will believe q.

Reasoners of Type 3. We will say that a reasoner is *normal* if for any proposition p, if he believes p, then he believes that he believes p. (If he believes p, then he also believes Bp.) By a reasoner of type 3, we shall mean a normal reasoner of type 2.

Reasoners of type 3 have one more stage of self-awareness than those of type 2.

Reasoners of Type 4. A normal reasoner doesn't necessarily know that he is normal. If a reasoner is normal, then for any proposition p, the proposition Bp⊃BBp is *true* (if he believes p, then he believes Bp), but the reasoner is not necessarily aware of the truth of Bp⊃BBp. Well, we will say that a reasoner *believes* that he is normal if for every proposition p, he believes the proposition Bp⊃BBp. (For every proposition p, the reasoner believes: "If I should ever believe p, then I will believe that I believe p.")

A reasoner of type 3 is in fact normal. By a reasoner of type 4, we mean a reasoner of type 3 who *knows* that he is normal. Thus for any proposition p, a reasoner of type 4 believes the proposition Bp⊃BBp.

As we have remarked, reasoners of type 4 play a major role in this book. Let us review the conditions defining a reasoner of type 4.

(1a) He believes all tautologies.

(1b) If he believes p and believes p⊃q, then he believes q.

(2) He believes (Bp&B(p⊃q))⊃Bq.

(3) If he believes p, then he believes Bp.

(4) He believes Bp⊃BBp.

SOME BASIC PROPERTIES OF SELF-AWARE REASONERS

We will now establish a few basic properties of reasoners of types 2, 3, and 4 that will be used throughout the remaining chapters.

First, a simple observation about reasoners of type 2: For any proposition p and q, the proposition $(Bp\&B(p\supset q))\supset Bq$ is logically equivalent to the proposition $B(p\supset q)\supset(Bp\supset Bq)$—because for *any* propositions X, Y, and Z, the proposition $(X\&Y)\supset Z$ is logically equivalent to $Y\supset(X\supset Z)$, as the reader can easily verify—and so any reasoner of type 2 believes all propositions of the form $B(p\supset q)\supset(Bp\supset Bq)$. Conversely, any reasoner of type 1 who believes all propositions of the form $B(p\supset q)\supset(Bp\supset Bq)$ must be of type 2. Let us record this as Fact 1.

Fact 1. A reasoner of type 1 is of type 2 if and only if he believes all propositions of the form $B(p\supset q)\supset(Bp\supset Bq)$.

Suppose now a reasoner of type 2 believes $B(p\supset q)$. He also believes $B(p\supset q)\supset(Bp\supset Bq)$, according to Fact 1, and being of type 1 (since he is of type 2) he will then believe $Bp\supset Bq$—which is a logical consequence of $B(p\supset q)$ and $B(p\supset q)\supset(Bp\supset Bq)$. And so as an obvious consequence of Fact 1 we have the following corollary: If a reasoner of type 2 believes $B(p\supset q)$, then he will believe $Bp\supset Bq$.

Now we come to some less obvious facts about reasoners of type 2.

1

Show that for any reasoner of type 2 and any propositions p, q, and r:

(a) He will believe $B(p\supset(q\supset r))\supset(Bp\supset(Bq\supset Br))$.

(b) If he ever believes $B(p\supset(q\supset r))$, then he will believe $Bp\supset(Bq\supset Br)$.

2

Show that if a reasoner of type 3 believes $p \supset (q \supset r)$, then he will believe $Bp \supset (Bq \supset Br)$. This fact will have many applications.

3 · Regularity

In the last chapter we defined a reasoner to be of type 1* if he is of type 1 and if for any propositions p and q, if he ever believes $p \supset q$, he will also believe $Bp \supset Bq$. This second condition will be given a name—we will call a reasoner *regular* if his belief in $p \supset q$ implies his belief in $Bp \supset Bq$.

Prove that every reasoner of type 3 is regular (and is thus of type 1*).

4

Prove that if a regular reasoner of type 1 believes $p \equiv q$, then he will believe $Bp \equiv Bq$.

5

There is an interesting connection between regularity and normality. For a regular reasoner of type 1, if there is so much as one proposition q such that he believes Bq, then he must be normal. Why is this?

6

Any reasoner of type 1 who believes p and believes q will believe p&q (which is a logical consequence of the two propositions p and q). Thus the proposition $(Bp\&Bq) \supset B(p\&q)$ is *true* for any reasoner of type 1, hence true for any reasoner of type 3.

Prove that any reasoner of type 3 *believes* (Bp&Bq)⊃B(p&q). (He *knows* that if he should ever believe p and believe q, he will believe p&q.)

CONSISTENCY

We say that a reasoner is *consistent* if the set of all propositions that he believes (or has believed or will believe) is a consistent set, and we shall say that he is *inconsistent* if his set of beliefs is inconsistent. For any reasoner of type 1, the set of his beliefs is logically closed; hence it follows from Principle C of Chapter 8 that the following three conditions are equivalent:

(1) He is inconsistent (he believes ⊥).
(2) He believes some proposition p and its negation (∼p).
(3) He believes all propositions.

We shall say that a reasoner *believes* he is consistent if he believes ∼B⊥ (he believes that he doesn't believe ⊥). We shall say that he *believes* that he is inconsistent if he believes B⊥ (he believes that he believes ⊥). A reasoner of even type 1 who is inconsistent will also believe that he is inconsistent (because he will believe everything), although a reasoner who believes that he is inconsistent is not necessarily inconsistent (although it can be shown that he must have at least one false belief).

Any reasoner of type 1 who believes some proposition p and its negation ∼p will be inconsistent—he will believe ⊥—and so the proposition (Bp&B∼p)⊃B⊥ is *true* for a reasoner of type 1.

7

Prove that any reasoner of type 3 *believes* the proposition (Bp&B∼p)⊃B⊥.

Note: This last problem is quite crucial for the next chapter. It means that for any proposition p, a reasoner of type 3 *knows* that if he should ever believe p and also believe ~p, then he will be inconsistent. (Of course this also applies to a reasoner of type 4, since every reasoner of type 4 is also of type 3.)

Inconsistency and Peculiarity. We recall that a reasoner is called peculiar if he believes some proposition p and also believes that he doesn't believe p. We have remarked that a peculiar reasoner is not necessarily inconsistent. However, any peculiar reasoner of type 3 *is* inconsistent, as the following problem will reveal.

8

Prove that any peculiar normal reasoner of type 1 must be inconsistent (and hence any peculiar reasoner of type 3 must be inconsistent).

Exercise 1. According to the above problem, for any proposition p, the proposition (Bp&B~Bp)⊃B⊥ is *true* for a reasoner of type 3, hence also true for a reasoner of type 4. Prove that any reasoner of type 4 correctly *believes* the proposition (Bp&B~Bp)⊃B⊥ (he *knows* that if he should ever be peculiar, he will be inconsistent).

9 · A Little Puzzle

Suppose a reasoner of type 4 believes p≡Bq. Will he necessarily believe p⊃Bp?

AWARENESS OF SELF-AWARENESS

Reasoners of type 4 have one marvelous property not shared by reasoners of lower types—namely, that they *know* they are of type

4, in a sense we will precisely define. Thus, for example, a reasoner may be of type 3 without knowing it, but a reasoner cannot be of type 4 unless he knows it.

Let us say that a reasoner *believes* he is of type 1 if he believes all propositions of the form BX, where X is any tautology, and believes all propositions of the form (Bp&B(p⊃q))⊃Bq. If he also believes all propositions of the form B((Bp&B(p⊃q))⊃Bq), then we will say that he *believes* he is of type 2. If he also believes all propositions of the form Bp⊃BBp, then we will say that he *believes* he is of type 3. If he also believes all propositions of the form B(Bp⊃BBp), then we will say that he *believes* he is of type 4. For each of these types, we will say that a reasoner *knows* that he is of that type if he believes he is of that type and really is of that type.

It is not difficult to see that a reasoner who knows that he is of type 1 is of type 2, and that any reasoner of type 3 knows that he is of type 2 (though he doesn't necessarily know that he is of type 3). Also a reasoner is of type 4 if and only if he knows that he is of type 3. The reader should try proving these facts as exercises.

The following problem is more interesting.

10

Prove that a reasoner of type 4 knows that he is of type 4.

This problem is interesting for several reasons. For one thing, it shows that being of type 4 constitutes a natural resting place in our hierarchy of reasoners. (It would be pointless, for example, to define a reasoner to be of type 5 if he is a reasoner of type 4 who knows that he is of type 4, since any reasoner of type 4 already knows he is of type 4, and thus we wouldn't get anything new.)

Secondly, anything that you or I can prove about all reasoners of type 4, using just propositional logic, any reasoner of type 4 can prove about himself, since he also knows propositional logic and knows that he is of type 4.

Exercise 2. To say that a reasoner is regular is to say that for any propositions p and q, the proposition $B(p{\supset}q){\supset}B(Bp{\supset}Bq)$ is *true* of the reasoner. (The proposition $B(p{\supset}q){\supset}B(Bp{\supset}Bq)$ is the proposition that if the reasoner believes p⊃q, then he will believe Bp⊃Bq.) Let us say that a reasoner *believes* that he is regular if he *believes* all propositions of the form $B(p{\supset}q){\supset}B(Bp{\supset}Bq)$.

Prove that every reasoner of type 4 knows that he is regular (i.e., he is regular and believes that he is regular).

Exercise 3. Prove that if a reasoner of type 4 believes $p{\supset}(Bp{\supset}q)$, then he will believe Bp⊃Bq. (The solution to this exercise will be given in Chapter 15; see page 126.)

SOLUTIONS

1 · Suppose the reasoner is of type 2. Take any propositions p, q, and r.

By Fact 1, he believes the following: (1) $B(p{\supset}(q{\supset}r)){\supset}(Bp{\supset}B(q{\supset}r))$. The reason is that for any propositions X and Y, he believes $B(X{\supset}Y){\supset}(BX{\supset}BY)$, so take p for X and (p⊃q) for Y.

Again, by Fact 1, he believes: (2) $B(q{\supset}r){\supset}(Bq{\supset}Br)$.

The following proposition is a logical consequence of (2):

(3) $(Bp{\supset}B(q{\supset}r)){\supset}(Bp{\supset}(Bq{\supset}Br))$. The reason is that, for any propositions X, Y, and Z, the proposition $(X{\supset}Y){\supset}(X{\supset}Z)$ is a logical consequence of Y⊃Z, as the reader can verify. We take Bp for X, B(q⊃r) for Y, and Bq⊃Br for Z, and we see that (3) is a logical consequence of (2).

We now know that the reasoner believes both (1) and (2), and $B(p{\supset}(q{\supset}r)){\supset}(Bp{\supset}(Bq{\supset}Br))$ is a logical consequence of (1) and (2), so the reasoner believes it. This proves (a).

(b) Suppose the reasoner believes $B(p{\supset}(q{\supset}r))$. By (a) he also believes $B(p{\supset}(q{\supset}r)){\supset}(Bp{\supset}(Bq{\supset}Br))$; hence he will believe Bp⊃(Bq⊃Br), since it is a logical consequence of the two propositions above.

Note: I will not give such detailed arguments in the future. By now the reader should have enough experience to follow briefer arguments and supply missing steps.

2 · Consider now a reasoner of type 3 who believes p⊃(q⊃r). Since he is normal, he will believe B(p⊃(q⊃r)). Then, by (b) of the last problem, he will believe Bp⊃(Bq⊃Br).

3 · Suppose a reasoner of type 3 believes p⊃q. Since he is normal, he will then believe B(p⊃q). Then, by the corollary to Fact 1, he will believe Bp⊃Bq. This proves that he is regular.

4 · Suppose a regular reasoner of type 1 believes p≡q. Then he will believe both p⊃q and q⊃p (since they are both logical consequences of p≡q). Being regular, he will then believe both Bp⊃Bq and Bq⊃Bp. Then, being of type 1, he will believe Bp≡Bq (which is a logical consequence of the last two propositions).

5 · Consider a regular reasoner of type 1. Suppose q is some proposition such that the reasoner believes Bq. Now, let p be any proposition that the reasoner believes. We are to show that he will believe Bp.

The proposition p⊃(q⊃p) is a tautology (as the reader can verify), hence the reasoner believes it. He also believes p (by assumption), hence he will believe q⊃p. Then, since he is regular, he will believe Bq⊃Bp. Then, since he believes Bq, he will believe Bp. This proves that he is normal.

6 · We are considering a reasoner of type 3. Now, the proposition p⊃(q⊃(p&q)) is obviously a tautology, hence the reasoner will believe it. Then, by (b) of Problem 1, he will believe Bp⊃(Bq⊃B(p&q)), hence he will believe the logically equivalent proposition (Bp&Bq)⊃B(p&q). (For any proposition X, Y, and Z, the proposition X⊃(Y⊃Z) is logically equivalent to (X&Y)⊃Z.)

7 · The proposition $(p\&\sim p)\supset\perp$ is a tautology, hence any reasoner of type 3 (or even of type 1) will believe it. Since a reasoner of type 3 is regular (by Problem 3), he will then believe $B(p\&\sim p)\supset B\perp$. He also believes $(Bp\&B\sim p)\supset B(p\&\sim p)$ (by Problem 6). Believing these last two propositions, he will believe $(Bp\&B\sim p)\supset B\perp$, which is a logical consequence of them.

Remarks. For *any* proposition q, the proposition $p\supset(\sim p\supset q)$ is a tautology. Hence by the above argument applied to q, instead of $\perp$, a reasoner of type 3 will believe $(Bp\&B\sim p)\supset Bq$.

8 · This is pretty obvious. If a normal reasoner believes p, he will believe Bp. If he also believes $\sim Bp$ and is of type 1, he will be inconsistent. Thus if a normal reasoner of type 1 believes p and believes $\sim Bp$, he will be inconsistent.

Solution to Exercise 1. A reasoner of type 4 believes $Bp\supset BBp$. He thus believes its logical consequence $(Bp\&B\sim Bp)\supset(BBp\&B\sim Bp)$. He also believes $(BBp\&B\sim Bp)\supset B\perp$ (by Problem 7, since $\sim Bp$ is the negation of Bp). The proposition $(Bp\&B\sim p)\supset B\perp$ is a logical consequence of the last two propositions, and so the reasoner will believe it.

In other words, a reasoner of type 4 can reason thus: "Suppose I ever believe p and also believe $\sim Bp$. Since I will believe p, I will also believe Bp, hence I will believe both Bp and $\sim Bp$, and then I will be inconsistent. And so, if I ever believe p and $\sim Bp$, I will be inconsistent."

9 · Suppose a reasoner of type 4 believes $p\equiv Bq$. He is regular (by Problem 3, since he is also of type 3), and so by Problem 3 he will believe $Bp\equiv BBq$. Hence he will believe $BBq\supset Bp$. He also believes $Bq\supset BBq$ (since he is of type 4), hence by propositional logic he will believe $Bq\supset Bp$. From $p\equiv Bq$ and $Bq\supset Bp$, he will deduce $p\supset Bp$. And so the answer is yes.

98

10 · Suppose the reasoner is of type 4. He then satisfies all the conditions that define a reasoner of type 4. We are to show that he *believes* all these conditions.

(1a) Take any tautology X. Being of type 4 (and hence of type 1), he believes X. Then, since he is normal, he believes BX. Thus for any tautology X, he believes BX.

(1b) It follows from the fact that he is of type 2 that he believes all propositions of the form $(Bp\&B(p \supset q)) \supset Bq$.

At this point we realize that he knows that he is of type 1. (In fact, by the above argument, any normal reasoner of type 2—i.e., any reasoner of type 3—knows that he is of type 1.)

(2) Since he believes all propositions of form $(Bp\&B(p \supset q)) \supset Bq$, and he is normal, then he believes $B((Bp\&B(p \supset q)) \supset Bq)$.

At this point we see that he knows he is of type 2. (In fact, any reasoner of type 3 knows that he is of type 2.)

(3) Since the reasoner is of type 4, then it is immediate that he knows all propositions of the form $Bp \supset BBp$—i.e., he knows that he is normal.

At this point we see that a reasoner of type 4 knows that he is of type 3. (But a reasoner of type 3 doesn't necessarily know that he is of type 3, because he may not know that he is normal.)

(4) Since the reasoner of type 4 believes $Bp \supset BBp$, and he is normal, he believes $B(Bp \supset BBp)$.

Now we see that a reasoner of type 4 knows that he is of type 4 (that is, he knows all of the propositions characterizing a reasoner of type 4).

Solution to Exercise 2. Consider a reasoner of type 4. Since he is of type 1, he believes $B(p \supset q) \supset (Bp \supset Bq)$. Since he is regular, he then believes $BB(p \supset q) \supset B(Bp \supset Bq)$. He also believes $B(p \supset q) \supset BB(p \supset q)$, since he knows he is normal. From this and the last fact, it follows that he must believe $B(p \supset q) \supset B(Bp \supset Bq)$.

· 12 ·

The Consistency Predicament

Now the stage is set, and the real show begins!

A reasoner of type 4 visits the Island of Knights and Knaves and believes the rules of the island. And so whenever a native makes a statement, the reasoner will believe that if the native is a knight, the statement is true, and its converse. Thus, if a native asserts a proposition p, then the reasoner will believe k⊃p (where k is the proposition that the native is a knight), and he will also believe p⊃k. Moreover, since the reasoner is of type 4, he is regular (as we showed in Problem 3 of the last chapter), and so if the native asserts p, the reasoner will believe not only k⊃p, but also Bk⊃Bp—that is, he will believe: "If I ever believe that he is a knight, then I will believe what he said."

We recall from the last chapter (Problem 7) that any reasoner of type 4, or even of type 3, knows that if he should ever believe p and believe ~p, he will be inconsistent (p can be any proposition).

Let us review and label these facts.

Fact 1. Suppose a native makes a statement to a reasoner of type 4. Then, (a) The reasoner will believe that if the native is a knight, the statement must be true (and conversely, that if the statement is true, then the native must be a knight). (b) The reasoner will also believe that if he should ever believe that the native is a knight, then he will believe what the native said.

Fact 2. For any proposition p, a reasoner of type 4 knows that if he should ever believe p and also believe ~p, then he will be inconsistent.

With these facts in mind, we are ready to embark. The first big result to which we turn is the following theorem.

Theorem 1 (after Gödel's Consistency Theorem). Suppose a native of the island says to a reasoner of type 4: "You will never believe that I am a knight." Then if the reasoner is consistent, he can never know that he is consistent; or, stated otherwise, if the reasoner ever believes that he cannot be inconsistent, he will become inconsistent!

1

Prove Theorem 1.

Solution. Suppose the reasoner does believe that he is (and will remain) consistent. We will show that he will become inconsistent.

The reasoner reasons: "Suppose I ever believe that the native is a knight. Then I'll believe what he said—I'll believe that I don't believe that he is a knight. But also, if I believe he's a knight, then I'll believe that I *do* believe he's a knight (since I am normal). Therefore, if I ever believe that he's a knight, then I'll believe both that I do believe he's a knight and that I don't believe he's a knight, which means I will be inconsistent. Now, I'll never be inconsistent [sic!], hence I will never believe he's a knight. He said that I would never believe he's a knight, and what he said was true, hence he is a knight."

At this point, the reasoner believes the native is a knight, and since he is normal, he will then know that he believes this. Hence the reasoner will continue: "Now I believe he is a knight. He said that I never would, hence he made a false statement, so he is not a knight."

101

At this point the reasoner believes that the native is a knight and also believes that the native is not a knight, and so he is now inconsistent.

Discussion. The important mathematical content of the above theorem can be presented without reference to knights and knaves. The function of the knight-knave island was to provide a simple method of getting some proposition k (in this case, that the native is a knight) such that the reasoner would believe "k is true if and only if I will never believe k." Any other method of getting such a proposition k would serve as well. Thus Theorem 1 is but a special case of the following theorem (which has nothing to do with knights and knaves).

Theorem G. If a consistent reasoner of type 4 believes some proposition of the form p≡~Bp, then the reasoner can never know that he is consistent. Stated otherwise, if a reasoner of type 4 believes p≡~Bp and believes that he is (and will remain) consistent, then he will become inconsistent.

We shall prove Theorem G in a sharper form.

Theorem G#. Suppose a normal reasoner of type 1 believes a proposition of the form p≡~Bp. Then:

(a) If he ever believes p, he will become inconsistent.

(b) If he is of type 4, then he knows that if he should ever believe p, then he will become inconsistent—i.e., he will believe the proposition Bp⊃B⊥.

(c) If he is of type 4 and believes that he cannot be inconsistent, then he will become inconsistent.

Proof. (a) Suppose he believes p. Being normal, he will then believe Bp. Also, since he believes p and believes p≡~Bp, he must believe ~Bp (since he is of type 1). And so he will then believe both Bp and ~Bp, hence he will be inconsistent.

102

(b) Suppose he is of type 4. Since he is of type 1 and believes $p\equiv\sim Bp$, he must also believe $p\supset\sim Bp$. Also, he is regular, hence he will then believe $Bp\supset B\sim Bp$. He also believes $Bp\supset BBp$ (since he knows he is normal). Hence he will believe $Bp\supset(BBp\&B\sim Bp)$, which is a logical consequence of the last two propositions. He also believes $(BBp\&B\sim Bp)\supset B\bot$ (by Fact 2, since for any proposition X, he believes $(BX\&B\sim X)\supset B\bot$, and so he believes this in the special case where X is Bp). Once he believes both $Bp\supset(BBp\&B\sim Bp)$ and $(BBp\&B\sim Bp)\supset B\bot$, he will have to believe $Bp\supset B\bot$ (since he is of type 1).

(c) Since he believes $Bp\supset B\bot$ (as we have just proved), then he also believes $\sim B\bot\supset\sim Bp$. Now, suppose he believes $\sim B\bot$ (he believes he cannot be inconsistent). Since he also believes $\sim B\bot\supset\sim Bp$ (as we have just seen), then he will believe $\sim Bp$. Since he also believes $p\equiv\sim Bp$, he will believe p, hence he will be inconsistent by (a).

The Student and His Theology Professor. Let us now turn again to the student and his theology professor who says to him: "God exists if and only if you will never believe that God exists." If the student believes the professor, then he believes the proposition $g\equiv\sim Bg$, where g is the proposition that God exists. Then, according to Theorem G, the student cannot believe in his own consistency without becoming inconsistent.

We hinted at this at the end of Chapter 2, but we were not then able to state what constituted a "reasonable" set of assumptions about the student's reasoning abilities. Now we can do this. The assumptions are simply that the student is a reasoner of type 4.

Of course the student has the option of believing in his own consistency without becoming inconsistent; he can simply refuse to believe the professor!

Exercise 1. Suppose that in Theorem 1 we are given the additional information that the rules of the island really do hold. Is the native a knight or a knave?

103

Exercise 2. In the example of the student and his theology professor, suppose that the student does believe the professor and also believes in his own consistency. If God really exists, then was the professor's statement true or false? If God doesn't exist, then was the professor's statement true or false?

Answer to Exercise 1. By Theorem 1, the reasoner will be inconsistent, hence will believe everything. In particular, he will believe that the native is a knight. Since the native said he wouldn't, then the native's statement was false. Therefore, if the rules of the island really hold, the native must be a knave.

Answer to Exercise 2. Again, the student will become inconsistent and believe everything. In particular, he will believe that God exists (he will also believe that God doesn't exist, but this is not relevant here). Letting g be the proposition that God exists, the proposition Bg is thus true, hence $\sim$Bg is false. If God does exist, then g is true, hence g$\equiv\sim$Bg is false, and the professor was wrong. If God doesn't exist, then g is false, g$\equiv\sim$Bg is true, and the professor was right.

Exercise 3. We have proved in earlier chapters the following three facts:

(1) If a native says to a reasoner of type 1*, "You will never believe I'm a knight," and the reasoner believes he will never be peculiar, then he will become peculiar. (Theorem 3, Chapter 10, page 84.)

(2) A peculiar reasoner of type 3 is inconsistent. (Problem 8, Chapter 11, page 94.)

(3) A reasoner of type 4 believes that if he should become peculiar, he will be inconsistent. (Exercise 1, Chapter 11, page 94.)

Using these three facts, one can give a much swifter proof of Theorem 1 of this chapter (page 101) than the one we have given. How?

Answer to Exercise 3. Suppose the native makes this statement— "You will never believe that I am a knight"—to a reasoner of type

104

4 and the reasoner believes that he will never be inconsistent. Then the reasoner will believe that he cannot be peculiar (because he knows that if he is peculiar, he will be inconsistent). Then by Fact 1 above, he will become peculiar. And by Fact 2 above, he will become inconsistent.

THE DUAL OF THEOREM G

Exercise 4. Suppose that the native, instead of saying, "You will never believe I'm a knight," says, "You *will* believe I'm a knave." Assuming the reasoner is of type 4 and believes in his own consistency, does it now follow that he will become inconsistent?

Solution. The answer is yes, and this can be easily established as a corollary of Theorem G, but first I'd like to sketch a direct argument. The reasoner reasons: "Suppose he's a knight. Then what he said is true, which means that I'll believe he is a knave. Once I believe he's a knave, I'll believe the *opposite* of what he said—I'll believe that I *don't* believe he's a knave. But if I believe he's a knave, I'll also believe that I *do* believe he's a knave (because I'm normal), and hence I'll be inconsistent. Since I will never be inconsistent, he can't be a knight after all; he must be a knave. Now I believe he's a knave. He said I would, and so he's a knight."

At this point the reasoner is inconsistent.

What we have just proved is a special case of the following "dual" of Theorem G.

Theorem G°. If a consistent reasoner of type 4 believes some proposition of the form p≡B∼p, then he can never know that he is consistent.

105

2

Theorem G° can be proved by "dualizing" the argument for Theorem G, but it can be obtained much more simply as a corollary of Theorem G. How?

Solution. Suppose the reasoner believes p≡B∼p. Then he will believe ∼p≡∼B∼p. Let q be the proposition ∼p. Then the reasoner believes q≡∼Bq, and so the reasoner does believe a proposition of the form p≡∼Bp (namely, q≡∼Bq), and so the result follows by Theorem G.

Exercise 5. Suppose that in Exercise 4 we add the assumption that the rules of the island really hold and that the reasoner *does* believe in his own consistency. Is the native a knight or a knave?

Exercise 6. Suppose a theology professor says to his student of type 4: "God exists if and only if you will believe that God doesn't exist." Suppose the student believes the professor and also believes in his own consistency. Prove that if the professor's statement is true, then God must exist. Prove that if the professor's statement is false, then God doesn't exist.

Exercise 7. Suppose a native tells a reasoner of type 4: "You will never believe I'm a knight" (or, alternatively, says: "You will believe I'm a knave"). Prove that the reasoner *knows* that if he should ever believe in his own consistency, he will become inconsistent. (The solution of this will follow easily from results to be proved in later chapters.)

· 13 ·

Gödelian
Systems

ALL THE results we have so far proved about reasoners and their beliefs are counterparts of metamathematical results about mathematical systems and the propositions provable in them. Before turning to these, let us summarize the most significant facts we have proved in the last few chapters.

Summary I. Suppose a reasoner believes the rules of the island and a native says to him: "You will never believe that I am a knight." Or, more generally, suppose a reasoner believes some proposition of the form $p \equiv \sim Bp$ (p is true if and only if I will never believe p). Then:

(1) For a reasoner of type 1, if he believes that he is always accurate, then he will become inaccurate; in fact, he will become peculiar.

(2) For a reasoner of type 1*, if he believes that he will never be peculiar, then he will become peculiar.

(3) For a reasoner of type 3, if he believes that he will never be peculiar, then he will become inconsistent.

(4) For a reasoner of type 4, if he believes that he will always be consistent, then he will become inconsistent.

All these facts, particularly (4), are related to important metamathematical facts, which we will discuss briefly.

The types of systems investigated by Kurt Gödel have the following features. First, there is a well-defined set of propositions expressible in the system; these will be called the propositions *of* the system. One of these propositions is ⊥ (logical falsehood), and for any proposition p and q of the system, the proposition (p⊃q) is also a proposition of the system. The logical connectives &, v, ∼, ≡ can all be defined from ⊃ and ⊥ in the manner explained in Chapter 7.

Second, the system—call it "S"—has various axioms and logical rules making certain propositions *provable* in the system. We thus have a well-defined subset of the set of propositions of the system—namely, the set of *provable* propositions of the system.

Third, for any proposition p of the system, the proposition that p is *provable* in the system is itself a proposition *of* the system (it may be true or false, and it may be provable in the system, or then again it may not). We let Bp be the proposition that p is provable in the system. (The symbol "B" for "provable" was introduced by Gödel; it stands for the German word *beweisbar*.) By a fortunate coincidence, we have the symbol "B" used in two closely related situations. When we speak of *reasoners,* Bp means that the reasoner believes p; when we speak of mathematical *systems,* Bp means that p is *provable* in the system.

We now define a system S to be of type 1, 1*, 2, 3, 4, in exactly the same way as we did for reasoners: If all tautologies are provable in S, and if for any propositions p and p⊃q both provable in S, q is also provable in S—if these two conditions hold—then we say that S is of type 1. If also Bp⊃Bq is provable in S whenever p⊃q is provable in S, then we say that S is of type 1*. If also all propositions of the form (Bp&B(p⊃q))⊃Bq are provable in S, then we say that S is of type 2. If also S is normal (i.e., Bp is provable whenever p is), then we say that S is of type 3. Lastly, if all propositions of the form Bp⊃BBp are provable in S, then we say that S is of type 4. Of course, all the results of Chapter 11 that we proved for *reasoners* hold good also for *systems* (where we now reinterpret "B" to mean *provable* rather than *believed*). As for the results of Chapters 10 and 12, we need another condition to which we now turn.

Gödelian Systems. Gödel made the remarkable discovery that each of the systems which he investigated had the property that there was a proposition p such that the proposition $p \equiv \sim Bp$ was provable in the system. Such systems we will call *Gödelian* systems.

The proposition $p \equiv \sim Bp$ is a very curious one. Here we have a proposition p equivalent to its own nonprovability in the system! The proposition p can be thought of as saying, "I am not provable in the system." How Gödel managed to find such a proposition need not concern us now, although it will be taken up in a much later chapter.

In analogy to systems, we might define a reasoner to be a Gödelian reasoner if there is at least one proposition p such that he believes the proposition $p \equiv \sim Bp$. Of course we have been studying Gödelian reasoners throughout the last chapter. (If a reasoner comes to the Island of Knights and Knaves and believes the rules of the island, and if a native says to him, "You will never believe that I am a knight," then the reasoner believes the proposition $k \equiv \sim Bk$, hence he becomes a Gödelian reasoner.)

Let us now say that a system can prove its own accuracy if it can prove all propositions of the form $Bp \supset p$. We will say that it can prove its own nonpeculiarity if it can prove all propositions of the form $\sim(Bp \& B \sim Bp)$. We will say that the system can prove its own consistency if it can prove the proposition $\sim B\perp$.

Let us now restate our opening summary in terms of systems, rather than reasoners.

Summary I*. (1) If a Gödelian system of type 1 can prove its own accuracy, then it is inaccurate—in fact, peculiar.

(2) If a Gödelian system of type 1* can prove its own nonpeculiarity, then it is peculiar.

(3) If a Gödelian system of type 3 can prove its own non-peculiarity, then it is inconsistent.

(4) If a Gödelian system of type 4 can prove its own consistency, then it is inconsistent.

Item (4) above is the really important one; it is a generalized form of Gödel's famous Second Incompleteness Theorem.

Discussion. In Gödel's original 1931 paper, he took a particular system (the system of *Principia Mathematica* of Whitehead and Russell) and showed that the system, if consistent, couldn't prove its own consistency. He stated that his method applied not only to this particular system, but to a wide variety of systems. Indeed, the method applies to all Gödelian systems of type 4, as we have just seen.

Another important Gödelian system of type 4 is the system known as "Arithmetic" (more completely, "First-Order Peano Arithmetic"). This is formalization of the theory of the ordinary whole numbers 0, 1, 2, . . . Since Arithmetic is a Gödelian system of type 4, it is subject to Gödel's Consistency Theorem; hence if Arithmetic is consistent (which it is, since only true propositions are provable in it), then it cannot prove its own consistency.

When the mathematician André Weil heard about this, he made the famous quip, "God exists, since Arithmetic is consistent; the Devil exists, since we cannot prove it."

This quip, though delightful, is actually misleading. It's not that we can't prove the consistency of Arithmetic; it is that Arithmetic can't prove the consistency of Arithmetic! *We* certainly can prove the consistency of Arithmetic, but our proof cannot be formalized in Arithmetic itself.

Indeed, there has been a good deal of popular misunderstanding concerning Gödel's Second Theorem (the Consistency Theorem); this has been partly due to the irresponsibility of some science reporters and other popularizers. (I, of course, am certainly all for popularization, providing the popularization is not inaccurate.) One particular popularizer wrote: "Gödel's theorem means that we can never know that Arithmetic is consistent." This is sheer nonsense. To see how silly it is, suppose it had turned out that Arithmetic could prove its own consistency—or, to be more realistic, suppose we take some other system that *can* prove its own consistency. Would this be any guarantee of the consistency of the system? Of course not. If the system were inconsistent, then, being of type 1, it could prove *anything*, including its own consistency! To trust the consistency of a system on the grounds that it can prove its own consistency would be as foolish as to trust the veracity of a person on the grounds that he claims to be always truthful. No, the fact that Arithmetic can't prove its own consistency doesn't cast the faintest ray of doubt on the consistency of Arithmetic.

As a matter of fact, in this book we will construct several Gödelian systems of type 4, and we will *prove* their consistency beyond a shadow of a doubt. Then, we will show that by virtue of their very consistency, the systems will be unable to prove their own consistency.

Exercise. Consider a system S of type 4 (but not necessarily Gödelian). Recall that for any proposition p, the proposition Bp is true if and only if p is provable in S.

(a) First show that for any proposition p, the proposition $(B(p \equiv \sim Bp) \& B \sim B\bot) \supset B\bot$ is true (for the system S). This is quite easy.

(b) Then show that $(B(p \equiv \sim Bp) \& B \sim B\bot) \supset B\bot$ is actually provable in S. (Not so easy!)

· 14 ·

More Consistency Predicaments

SOME PRELIMINARY PROBLEMS

1

Suppose a reasoner believes that he is inconsistent. Is he necessarily inconsistent? Is he necessarily inaccurate?

2

Suppose a reasoner believes that he is inaccurate. Prove that he is right!

3

Suppose a reasoner of type 1* believes that he cannot be inconsistent (he believes $\sim B\bot$). Will he necessarily believe that he will never believe that he is inconsistent? (I.e., will he necessarily believe $\sim BB\bot$?)

SOME MORE CONSISTENCY PREDICAMENTS

We are back to the Island of Knights and Knaves. We continue to assume that the reasoner is of type 4 and that he believes the rules of the island.

4

Suppose a native says to the reasoner, "If I am a knight, then you will believe that I'm a knave." Prove:

(a) The reasoner will sooner or later believe himself inconsistent.

(b) If the rules of the island really hold, then the reasoner will become inconsistent!

Note: In the problems of this chapter, we are *not* assuming that the reasoner believes that he is consistent.

5

Suppose the native says instead, "If I am a knight, then you will never believe that I am one." Prove:

(a) The reasoner will become inconsistent.

(b) The rules of the island don't really hold.

Note 1: This problem holds good even if the reasoner is only of type 3.

Note 2: For a "student and his theology professor" version of the last two problems, see the discussion following the solution to Problem 5.

6

Here is a curious one. A reasoner of type 4 comes to what he *believes* is a knight-knave island (he believes the rules of the island), and a native says to him the following two things:

(1) "You will believe that I am a knave."

(2) "You will always be consistent."

Prove that the reasoner will become inconsistent and that the rules of the island don't hold.

7

This time a native makes the following two statements to a reasoner of type 4:

(1) "You will never believe I'm a knight."

(2) "If you ever believe I'm a knight, then you will become inconsistent."

Prove that the reasoner will become inconsistent and that the rules of the island don't hold.

TIMID REASONERS

Let us say that a reasoner shouldn't believe p if his believing p will lead him into an inconsistency. Let us say that a reasoner is *afraid* to believe p if he believes $Bp \supset B\bot$—i.e., if he believes that his believing p will lead him into an inconsistency. (In other words, he is afraid to believe p if he believes that he shouldn't believe p.)

We know that any reasoner of type 4 who believes the rules of the island and is told by a native that he will never believe that he is a knight *shouldn't* believe in his own consistency. In general, however, there is no reason why a reasoner of type 4 shouldn't believe in his own consistency. But now arises a very curious thing: If for some reason, a reasoner of type 4 *is* afraid of believing in his own consistency, his very fear justifies it. By this I mean that any

114

reasoner of type 4 who is afraid of believing in his own consistency, really *shouldn't* believe in his own consistency. Put still another way, if a reasoner of type 4 believes that his believing in his own consistency will get him into an inconsistency, then it will!

Surprising as this fact may be, it is not difficult to prove. Moreover, this fact holds even for normal reasoners of type 1.

8

Prove that if a normal reasoner of type 1 is afraid to believe in his own consistency, then he really shouldn't believe in his own consistency.

Remarks. In the above problem, we see that for a normal reasoner of type 1, his belief in the proposition $B{\sim}B{\perp}{\supset}B{\perp}$ is *self-fulfilling* in the sense that his believing that proposition is a sufficient condition for the proposition being true. The theme of self-fulfilling beliefs will play a major role in the next few chapters.

9

A reasoner of type 4 is also a normal reasoner of type 1, hence according to the last problem, if he is afraid to believe in his own consistency, then he shouldn't believe in his own consistency. This means that for a reasoner of type 4, the proposition $B(B{\sim}B{\perp}{\supset}B{\perp}){\supset}(B{\sim}B{\perp}{\supset}B{\perp})$ is true. Prove that any reasoner of type 4 *knows* that this proposition is true (he knows that if he is afraid to believe in his own consistency, then he shouldn't believe in it).

10

Suppose a reasoner of type 4 *believes* that he is afraid to believe in his own consistency. Does it follow that he really *is* afraid to believe in his own consistency?

SOLUTIONS

1 · Suppose a reasoner believes that he is inconsistent. I see no reason why he is necessarily inconsistent, but he must be inaccurate for the following reasons.

A reasoner who believes he is inconsistent is either right or wrong in this belief. If he is wrong in this belief, then he is obviously inaccurate (he has the false belief that he is inconsistent). If he is right in this belief, then he really does believe the false proposition ⊥. In either case, he has at least one false belief.

2 · If he were wrong, then he would be accurate, which is a contradiction.

3 · Suppose he believes ~B⊥. Then he believes the logically equivalent proposition B⊥⊃⊥. Also he is regular (since he is of type 1*), and hence he will believe BB⊥⊃B⊥, hence he will believe the logically equivalent proposition ~B⊥⊃~BB⊥. Since he also believes ~B⊥, then he will believe ~BB⊥.

4 · We know by Theorem 1 of Chapter 3 that for any proposition p, if a native of a knight-knave island says, "If I am a knight then p," the native must be a knight and p must be true. Now, any reasoner—even if type 1—knows this as well as you and I, and so if the reasoner *believes* that the rules of the island hold, then if a native says to him, "If I am a knight, then p," the reasoner will believe that the native is a knight and that p is true. In this particular problem, the native has said, "If I am a knight, then you will believe that I am a knave," and so the reasoner will believe that the native is a knight and also that he (the reasoner) will *believe* that the native is a knave. And so the reasoner will believe k and also believe B~k (k is the proposition that the native is a knight). So far, we have used only the fact that the reasoner is of type 1. However, he is of type 4, and since he believes k, he will also believe Bk. Hence he will

116

believe Bk and believe B~k, but he knows that (Bk&B~k)⊃B⊥ (as we showed in Chapter 11, page 98). Therefore he will believe B⊥—i.e., he will believe that he is (or will be) inconsistent. He also believes that the native is a knight.

So far, we have not used the fact that the rules of the island really hold; our preceding argument used only the fact that the reasoner *believes* that the rules hold. Now, suppose the rules really do hold. Then the native really is a knight and the reasoner really will believe that the native is a knave (according to Theorem 1, Chapter 3). But since the reasoner also believes that the native is a knight, he will become inconsistent.

5 · This time the reasoner believes k≡(k⊃~Bk), hence he believes k and believes ~Bk, which are logical consequences of k≡(k⊃~Bk). Since he believes k, he will believe Bk, and believing ~Bk, he will become inconsistent.

If the rules of the island really hold, then k≡(k⊃~Bk) is not only believed by the reasoner, but is actually true, hence k&~Bk (which is logically implied by it) is true, hence ~Bk is true, contrary to the fact that the reasoner *does* believe the native is a knight (and hence Bk, rather than ~Bk, is true). Thus the rules of the island don't really hold.

Discussion. Let us look at the last two problems in the form of the student and his theology professor. Suppose the professor says: "God exists, but you will never believe that God exists." If the student is of type 4 and believes the professor, he will have to believe himself inconsistent. If also the professor's statement was true, then the student really will become inconsistent.

On the other hand, suppose the professor says: "God exists, but you will never believe that God exists." Then if the student is of type 4—or even of type 3—and believes the professor, he will become inconsistent, and also, the professor's statement is false.

117

6 · The reasoner reasons: "Suppose he is a knave. Then his second statement is false, which means that I will be inconsistent, hence I'll believe everything—in particular, that he's a knave. But this will validate his first statement and make him a knight. Therefore it is contradictory to assume that he is a knave, hence he must be a knight. Since he is a knight, his first statement is true, hence I will believe he is a knave. But I now believe he is a knight, hence I will be inconsistent. This proves that I will be inconsistent. However, he's a knight and said that I will always be consistent. Therefore I won't ever be inconsistent."

At this point the reasoner has come to the conclusion that he will become inconsistent and that he won't become inconsistent, and so he is now inconsistent.

Since the reasoner has become inconsistent, the native's second statement is false. Also, since the reasoner has become inconsistent, he will believe everything, including the fact that the native is a knave. This makes the native's first statement true. Since the native has made one true statement and one false statement, then the rules of the island don't really hold.

7 · The reasoner believes the following two statements:

(1) $k \equiv \sim Bk$
(2) $k \equiv (Bk \supset B\bot)$

Since he believes (1), then, according to (b) of Theorem $G^\#$, Chapter 12 (page 102), the reasoner will believe $Bk \supset B\bot$. And, since he believes (2), he will believe k. So, according to (a) of Theorem $G^\#$, Chapter 12, he will become inconsistent.

It further follows that the native's first statement was false and his second statement true. Therefore the rules of the island don't hold.

8 · We assume that the reasoner is normal and of type 1 and that he believes $B \sim B\bot \supset B\bot$. We are to show that if he believes he is consistent, he will become inconsistent (and therefore that his fear

118

is justified). So, suppose he ever does believe $\sim B\perp$. Being normal, he will then believe $B\sim B\perp$. This, together with his belief in $B\sim B\perp \supset B\perp$, will cause him to believe $B\perp$ (because he is of type 1). And so he will believe both $B\perp$ and $\sim B\perp$ (he will believe that he is inconsistent and that he is not inconsistent), which will make him inconsistent.

9 · The reasoner reasons: "Suppose I become afraid to believe in my own consistency. This means that I'll believe that I shouldn't believe in my own consistency—i.e., I'll believe $B\sim B\perp \supset B\perp$. Since I am of type 1 (being of type 4), then I will believe $\sim B\perp \supset \sim B\sim B\perp$. Now suppose I should also believe that I am consistent—i.e., suppose I believe $\sim B\perp$. Then, since I will believe $\sim B\perp \supset \sim B\sim B\perp$, I'll believe $\sim B\sim B\perp$. But I'll also believe $B\sim B\perp$ (since I'll believe $\sim B\perp$ and I am normal), and hence I'll be inconsistent. Therefore, if I am afraid to believe in my own consistency, I really cannot believe in my own consistency without becoming inconsistent. Thus the proposition $B(B\sim B\perp \supset B\perp) \supset (B\sim B\perp \supset B\perp)$ is true."

10 · We have just seen that the reasoner *believes* the proposition $B(B\sim B\perp \supset B\perp) \supset (B\sim B\perp \supset B\perp)$. Therefore, if he believes $B(B\sim B\perp \supset B\perp)$, he will believe $B\sim B\perp \supset B\perp$—which means that if he *believes* that he is afraid to believe in his own consistency, then he really will be afraid to believe in his own consistency.

119

Part VI

SELF-FULFILLING BELIEFS AND LÖB'S THEOREM

· 15 ·

Self-Fulfilling
Beliefs

THE PROBLEMS of this chapter are all related to Löb's Theorem, a famous result that is important to the thrust of this book.

We now have a change of scenario: A reasoner of type 4 is thinking of visiting the Island of Knights and Knaves because he has heard a rumor that the sulfur baths and mineral waters there might cure his rheumatism. Before embarking, however, he discusses the situation with his family physician. He asks the doctor whether the "cure" really works. The doctor replies: "The cure is largely psychological; the belief that it works is self-fulfilling. *If you believe that the cure will work, then it will work.*"

The reasoner trusts his doctor implicitly, and so he goes to the island with the prior belief that if he believes the cure will work, then the cure will work. He takes the cure, which lasts only a day but which is not supposed to work for several weeks, if it works at all. The next day he starts worrying about the situation and thinks: "If only I can *believe* that the cure works, then it will work. But how do I know whether I will ever believe that it works? I have no rational evidence that the cure will work, nor do I have any evidence that I will ever believe that the cure will work. For all I know, I may never believe that the cure will work, and the cure might accordingly not work!"

A native of the island passes by and asks the reasoner why he looks so disconsolate. The reasoner explains the entire situation, then summarizes it by saying: "If I ever believe the cure will work, then it will, but will I ever believe the cure will work?" The native replies: *"If you ever believe I'm a knight, then you will believe that the cure will work."*

At first, this did not seem particularly reassuring to the reasoner. He thinks: "What good does this do me? Even if what he says is true, this will only reduce the problem to whether I will ever believe that he is a knight. How do I know whether I will ever believe he is a knight? And even if I do, he may be a knave and his statement may be false, hence I may still not believe that the cure will work." But then the reasoner thought some more about his problem, and after a while he heaved a sigh of relief. Why?

Well, as we will see, the amazing thing is that the reasoner *will* believe that the cure will work and, assuming that the doctor is right, the cure will work! I might remark that the rules of the island don't have to really hold for the argument to go through; it is enough that the reasoner *believes* that they hold.

This problem is closely related to M. H. Löb's important theorem, which we will examine later. But first, let's consider a slightly simpler problem, one that comes even closer to Löb's original argument. Suppose that the native, instead of saying the above, says: "If you ever believe that I'm a knight, then the cure will work."

1 · (After Löb)

Prove that under the above conditions, the reasoner *will* believe that the cure will work (and hence, if the doctor was right, then the cure will work).

Solution. It will be easiest to give the solution partly in words and partly in symbols. We let k be the proposition that the native is a knight, and we let C be the proposition that the cure will work. At the outset, the reasoner believes the proposition BC⊃C.

The reasoner reasons: "Suppose I ever believe that he is a knight. Then I'll believe what he says—I'll believe that Bk⊃C. Also, if I ever believe he's a knight, I'll believe that I believe he's a knight—I'll believe Bk. And so, if I ever believe he is a knight, I'll believe both Bk and Bk⊃C, hence I'll believe C. Thus, if I ever believe he is a knight, then I'll believe that the cure works. But if I ever believe that the cure works, then the cure will work (as my doctor told me). And so, if I ever believe he's a knight, then the cure will work. Well, that's exactly what he said. He said that if I ever believe he's a knight, then the cure will work, and he was right! Hence he is a knight."

At this point the reasoner believes that the native is a knight, and since the reasoner is normal, he continues: "Now I believe he is a knight. I have already proved that if I believe he is a knight, then the cure will work, and since I do believe that he is a knight, the cure will work."

At this point the reasoner believes that the cure will work. Then, assuming his doctor was right, the cure will work.

The solution to Problem 1 could have been established more quickly had we first proved the following lemma, which will have other applications as well.

Lemma 1. Given any proposition p, suppose a native of the island says to reasoner of type 4: "If you ever believe that I'm a knight, then p is true." Then the reasoner will believe: "If I ever believe he is a knight, then I will *believe* p." More generally, for any two propositions k and p, if a reasoner of type 4 believes the proposition k≡(Bk⊃p), or even believes the weaker proposition k⊃(Bk⊃p), then he will believe Bk⊃Bp.

Exercise 1. How is Lemma 1 proved? (This is the same problem found in Exercise 3, Chapter 11, page 96.)

Exercise 2. How does the use of Lemma 1 facilitate the solution of Problem 1?

Answer to Exercise 1. Let's first show this in the knight-knave version. We let k be the proposition that the native is a knight. The native has asserted the proposition Bk⊃p. The reasoner reasons: "If I ever believe that he's a knight, then I'll believe what he says—I'll believe Bk⊃p. But if I believe he's a knight, I'll also believe Bk (I'll believe that I believe he's a knight). Once I believe Bk⊃p and I believe Bk, then I'll believe p. And so, if I ever believe he's a knight, then I'll believe p."

Of course the more general form can be proved in essentially the same manner, or alternatively as follows: Suppose a reasoner of type 4 believes k⊃(Bk⊃p), which he will certainly believe, if he believes the stronger proposition k≡(Bk⊃p). Then, according to Problem 2, Chapter 11, he will believe Bk⊃(BBk⊃Bp). He also believes Bk⊃BBk. Believing these two propositions, he will believe Bk⊃Bp, which is a logical consequence of them. (For any proposition X, Y, and Z, the proposition X⊃Z is a logical consequence of X⊃(Y⊃Z) and X⊃Y. In particular, this is so if X is the proposition Bk, Y is the proposition BBk, and Z is the proposition Bp.)

Answer to Exercise 2. The reasoner believes k≡(Bk⊃C)—because the native asserted Bk⊃C. Then, according to Lemma 1, the reasoner will believe Bk⊃BC. He also believes BC⊃C, hence he will believe Bk⊃C. Then he will believe k (since he believes both Bk⊃C and k≡(Bk⊃C)), hence he will believe Bk (he is normal). Now that he believes Bk and believes Bk⊃C, he will believe C.

The upshot of Problem 1 is the following theorem.

Theorem 1 (After Löb). For any proposition k and C, if a reasoner of type 4 believes BC⊃C and believes k≡(Bk⊃C), then he will believe C.

Theorem 1 yields the following curious result.

126

Exercise 3. Suppose a theology student is worried about such things as the existence of God and his own salvation. He asks his professor: "Does God exist?" And "Will I be saved?" The professor then makes the following statements:

(1) "If you believe that you will be saved, then you will be saved."

(2) "If God exists and you believe that God exists, then you will be saved."

(3) "If God doesn't exist, then you will believe that God exists."

(4) "You will be saved only if God exists."

Assuming that the student is a reasoner of type 4 and that he believes his professor, prove: (a) The student will believe that he will be saved; (b) If the professor's statements are true, then the student will be saved.

Solution. Let g be the proposition that God exists and let S be the proposition that the student will be saved. The student then believes the following four propositions:

(1) $BS \supset S$
(2) $(g \& Bg) \supset S$
(3) $\sim g \supset Bg$
(4) $S \supset g$

The proposition $\sim Bg \supset g$ is a logical consequence of (3). This proposition, together with (4), has as a logical consequence the proposition $(\sim Bg v S) \supset g$. Also $\sim Bg v S$ is logically equivalent to $Bg \supset S$, hence $(\sim Bg v S) \supset g$ is logically equivalent to $(Bg \supset S) \supset g$. Also $g \supset (Bg \supset S)$ is logically equivalent to (2), and $g \equiv (Bg \supset S)$ is a logical consequence of $(Bg \supset S) \supset g$ and $g \supset (Bg \supset S)$. Therefore $g \equiv (Bg \supset S)$ is a logical consequence of (1), (2), and (3). Since the student believes (1), (2), and (3), he will also believe $g \equiv (Bg \supset S)$. Since by (1) he also believes $BS \supset S$, then by Theorem 1, he will believe S. Thus BS is true, and if the professor was right, $BS \supset S$ is true, hence S is true, which means that the student will be saved.

. . .

Now let us return to Problem 1 and its proposition: "If you ever believe I'm a knight, then the cure will work."

2

Suppose a reasoner of type 4 again believes BC⊃C, but this time the native says: "If you ever believe I'm a knight, then you will *believe* that the cure works." Prove that the reasoner will again believe that the cure will work.

Solution. This time the native is asserting Bk⊃BC (instead of Bk⊃C). Then by Lemma 1, the reasoner will believe Bk⊃BBC (instead of Bk⊃BC). However, the reasoner believes BC⊃C, and since he is of type 4, he is regular, hence he will believe BBC⊃BC. Believing this and Bk⊃BBC, he will believe Bk⊃BC. He also believes k≡(Bk⊃BC), so he will believe k. Then he will believe Bk, and since he will believe Bk⊃BC, he will believe BC. But he also believes BC⊃C, hence he will believe C.

Of course the above argument generalizes as follows:

Theorem 2. Given any propositions k and C, if a reasoner of type 4 believes BC⊃C and believes k≡(Bk⊃BC), then he will believe C.

3

Again a reasoner of type 4 has the prior belief that if he believes that the cure will work, then the cure will work. This time the native says to him: "Sooner or later you will believe that if I am a knight, then the cure will work." We will see that again the reasoner will believe that the cure will work.

More generally, we will prove the following theorem.

Theorem 3. For any propositions k and C, if a reasoner of type 4 believes BC⊃C and believes k≡B(k⊃C), then he will believe C.

The proof of Theorem 3 is facilitated by the following two lemmas, which are of interest in their own right.

Lemma 2. Suppose that for some proposition q, a native says to a reasoner of type 4: "You will believe q." Then the reasoner will believe: "If he is a knight, then I will believe that if he is a knight, I will believe that he is a knight." (Stated more abstractly, for any propositions k and q, if a reasoner of type 4 believes k≡Bq, then he will believe k⊃Bk.)

Lemma 3. Given any proposition p, suppose a native says to a reasoner of type 4: "You will believe that if I am a knight, then p is true." Then the reasoner will believe: "If he is a knight, then I will *believe* p." (Stated more abstractly, if a reasoner of type 4 believes k≡B(k⊃p), then he will believe k⊃Bp.)

How are Lemmas 2 and 3 and Theorem 3 proved?

Proof of Lemma 2. This is Problem 9 of Chapter 11, which we have already solved, but I wish to give a knight-knave version of the proof, which is particularly intuitive. The native has said: "You will believe q." The reasoner then reasons: "Suppose he is a knight. Then I *will* believe q. Then I'll believe that I believe q, hence I'll believe what he said, hence I'll believe he's a knight. Therefore, if he is a knight, then I'll believe he's a knight."

Proof of Lemma 3. We shall use Lemma 2 to facilitate this proof.

The native has said: "You will believe that if I'm a knight, then p." Let q be the proposition "If I'm a knight, then p." The native has told the reasoner that he will believe q, and so by Lemma 2, the reasoner will believe that if the native is a knight, then he (the reasoner) will believe that the native is a knight. And so the reasoner reasons: "Suppose he is a knight. Then I'll believe that he is a knight. Then I'll believe what he says—I'll believe Bk⊃p. I'll also believe Bk (I'll believe that I believe he is a knight). Therefore, if he is a knight, then I'll believe Bk⊃p and I'll believe Bk, hence I will also believe p. And so if he is a knight, then I will believe p."

Proof of Theorem 3. I will give a knight-knave version of the proof. The native has said: "You will believe that if I am a knight, then the cure will work." By Lemma 3, the reasoner will believe: "If he is a knight, then I will believe that the cure will work." The reasoner then continues: "Also, if I believe that the cure will work, then the cure will work. Therefore, if he is a knight, the cure will work. I now believe that if he is a knight, then the cure will work. He said I would believe that, hence he is a knight. And so he is a knight, and in addition (as I have proved), if he is a knight, then the cure will work. Therefore the cure will work."

At this point the reasoner will believe that the cure will work.

Discussion. One can also obtain Theorem 3 as an easy corollary of Theorem 1 by the following argument. Suppose we are given propositions k and C such that a reasoner of type 4 believes k≡B(k⊃C) and believes BC⊃C. We are to show that he will believe C. Since he believes k≡B(k⊃C), he must also believe (k⊃C)≡(B(k⊃C)⊃C)), which is a logical consequence of k≡B(k⊃C)). Then he believes the proposition k′≡(Bk′⊃C), where k′ is the proposition k⊃C. Since he also believes BC⊃C, then he will believe C by Theorem 1.

The following curious exercise illustrates self-reference carried to its extreme.

Exercise 4. Suppose a native of the island says to a reasoner of type 4: "Sooner or later you will believe that if I am a knight, then you will believe that I am one."

(a) Prove that the reasoner will believe that the native is a knight.

(b) Prove that if the rules of the island really hold, then the native is a knight.

Solution. This is an easy consequence of Lemma 2.

(a) The native has asserted B(k⊃Bk). Thus there is a proposition q—namely, k⊃Bk—such that the native has claimed that the reasoner will believe q. Then by Lemma 2, the reasoner will believe k⊃Bk. Then, since he is normal, he will believe B(k⊃Bk)—he will believe what the native said. Hence he will believe that the native is a knight.

(b) Since the reasoner believes Bk, he certainly believes k⊃Bk, hence the native's statement was true. And so the native is a knight (if the rules of the island really hold).

The essential mathematical content of the above exercise is that for any proposition k, if a reasoner of type 4 believes k≡B(k⊃Bk), then he will also believe k. If also k≡B(k⊃Bk) is true, so is k.

DUAL FORMS

Problems 1, 2, and 3 (more generally, Theorems 1, 2, and 3) have their "dual" forms, which are rather curious.

1° · (Dual of Problem 1)

Again, a reasoner of type 4 comes to the island already believing that if he believes that the cure will work, then the cure will work. He meets a native who says to him: "The cure will not work and you will believe that I am a knave."

Prove that the reasoner will believe that the cure will work.

2° · (Dual of Problem 2)

Like Problem 1°, except that the native says: "You will believe that I'm a knave, but you will never believe that the cure will work." Show that the same conclusion follows (the reasoner will believe that the cure will work).

3° · (Dual of Problem 3)

This time the native says: "You will never believe that if I am a knave, then the cure will work." (Alternatively, he could say: "You will never believe that either I am a knight or that the cure will work.") Show that the same conclusion follows.

Solution to Problem 1°. We could prove this from scratch, but it is easier to take advantage of Theorem 1, which we have already proved.

The native has asserted ($\sim$C&B$\sim$k), and so the reasoner believes k$\equiv$($\sim$C&B$\sim$k). But we know k$\equiv$($\sim$C&B$\sim$k) is logically equivalent to $\sim$k$\equiv$$\sim$($\sim$C&B$\sim$k), which in turn is logically equivalent to $\sim$k$\equiv$(B$\sim$k$\supset$C). Therefore the reasoner believes $\sim$k$\equiv$(B$\sim$k$\supset$C), and so he believes a proposition of the form p$\equiv$(Bp$\supset$C)—p is the proposition $\sim$k—and so, by Theorem 1, if he believes BC$\supset$C, he will believe C.

The solutions of Problems 2° and 3° can likewise be obtained as corollaries of Theorems 2 and 3, respectively. We leave the verification to the reader.

Exercise 5. Suppose a native says to a reasoner of type 4: "If you ever believe I'm a knight, then you will be inconsistent." Is it possible for the reasoner to believe in his own consistency without becoming inconsistent? (Hint: Use Theorem 2.)

Exercise 6. Suppose a reasoner of type 4 believes that if he believes the cure will work, then it will work. Suppose we now have a native who says to him: "If you ever believe that you will believe I'm a knight, then the cure will work."

Will the reasoner necessarily believe that the cure will work?

Exercise 7. Suppose the native instead says: "You will believe that if you ever believe I'm a knight, then the cure will work."

Will the reasoner necessarily believe that the cure will work?

Exercise 8. The following dialogue ensues between a student and his theology professor:

> STUDENT: If I believe that God exists, then will I also believe that I will be saved?
> PROFESSOR: If that is true, then God exists.
> STUDENT: If I believe that God exists, then *will* I be saved?
> PROFESSOR: If God exists, then that is true.

Prove that if the professor is accurate and if the student believes the professor, then God must exist and the student will be saved.

Exercise 9. The following strengthening of Theorem 3 can be proved. A reasoner of type 4 comes to the island for the cure and has the prior belief that if he should ever believe that the cure will work, then it will. He asks a native: "Will I ever believe that if you are a knight, then the cure will work?" The native replies: "If that is not so, then the cure will work." (Alternatively, he could have replied: "Either that is so or the cure will work.") The problem is to prove that the reasoner will believe that the cure will work. (Stated more abstractly, if a reasoner of type 4 believes $k \equiv (CvB(k \supset C))$ and believes $BC \supset C$, then he will believe C.) The proof of this is facilitated by first proving the following two facts as lemmas:

(1) For any propositions p and q, suppose a native says to a reasoner of type 4: "Either p is true or you will believe q." Then the reasoner will believe: "If the native is a knight, then I will believe that either p is true, or that the native is a knight."

(2) For any propositions p and q, suppose a native says to a reasoner of type 4: "Either p is true or else you will believe that if I am a knight, then q is true." The reasoner will then believe: "If the native is a knight, then either p is true or I will believe that q is true."

133

Exercise 10. The last exercise has the following dual. Again, a reasoner of type 4 believes that if he should believe that the cure will work, then it will. He now meets a native who says: "The cure doesn't work and you will never believe that either I'm a knight or that the cure works." Prove that the reasoner will believe that the cure will work.

· 16 ·

The Rajah's Diamond

THOSE OF you who have read the magnificent story "The Rajah's Diamond," by Robert Louis Stevenson, will recall that at the end, the diamond was thrown into the Thames. Recent research, however, has revealed that the diamond was subsequently found by an inhabitant of the Island of Knights and Knaves who was vacationing in England at the time. One rumor has it that he then took the diamond to Paris and died shortly after. According to another, he took the diamond back home. If the second version is correct, then the diamond is somewhere on the knight-knave island.

A reasoner of type 4 decided to follow up on the second rumor in hopes of finding the diamond. He reached the island and believed the rules of the island. Also, the rules of the island really held. There are five different versions of what actually happened; each is of interest, and so I will relate them all.

1 · The First Version

According to the first version, when the reasoner reached the island, he met a native who made the following two statements:

(1) "If you ever believe that I am a knight, then you will believe that the diamond is on this island."

(2) "If you ever believe that I am a knight, then the diamond *is* on this island."

If this version is the correct one, is the diamond on the island?

2 · The Second Version

According to a second, slightly different version, the native, instead of making the two statements reported above, made the following two statements:

(1) "If I am a knave and if you ever believe that I'm a knight, then you will believe that the diamond is on this island."

(2) "I am actually a knight, and if you ever believe this, then the diamond is on this island."

If this second version is correct, is there sufficient evidence to conclude that the diamond must be on the island?

3 · The Third Version

The third version is particularly curious. According to it, the native made the following two statements:

(1) "If you ever believe that I'm a knight, then the diamond is not on this island."

(2) "If you ever believe that the diamond is on this island, then you will become inconsistent."

If this third version is correct, what conclusion should be drawn?

4 · The Fourth Version

According to the fourth version, the native made only one statement:

(1) "If you ever believe I'm a knight, then you will believe that the diamond is on this island."

The reasoner, of course, could get nowhere. He then discussed the whole affair with the Island Sage, who was known to be a knight of the highest integrity. The Sage made the following statement:

"If the native you spoke to is a knight and if you ever believe that the diamond is on this island, then the diamond is on this island."

If this fourth version is correct, what conclusion should be drawn?

5 · The Fifth Version

This is like the last version, except that the native said:

(1) "You will believe that if I am a knight, then the diamond is on this island."

The Sage made the same statement.

Assuming that these five versions are equally probable, what is the probability that the Rajah's diamond is on the Island of Knights and Knaves?

Remarks. The mathematical content of the last two problems constitutes strengthenings of Theorems 2 and 3 of the last chapter— see the discussion following the solutions.

SOLUTIONS

We let k be the proposition that the native is a knight. We let D be the proposition that the diamond is on the island.

1 · Since the native made the two statements which he made, then the reasoner will believe the following two propositions:

(1) $k \equiv (Bk \supset BD)$
(2) $k \equiv (Bk \supset D)$

And the reasoner will therefore certainly believe the following two weaker propositions:

(1)' $(Bk \supset BD) \supset k$
(2)' $k \supset (Bk \supset D)$

137

As we will see, the fact that the reasoner believes even (1)′ and (2)′ is enough to solve the problem.

Since he believes (2)′, then by Lemma 1 of the last chapter, he will believe Bk⊃BD. Believing this and believing (1)′, he will then believe k. Believing k and believing (2)′, he will then believe Bk⊃D. Also, since he believes k, he will believe Bk, and hence he will believe D. Therefore BD is true. Hence Bk⊃BD is true, and since (1) is true (the rules of the island really hold), then k is true (the native is really a knight). Then since (2) is true, the proposition Bk⊃D is true. Also Bk is true (we have seen that the reasoner will believe k), and thus D is true. Therefore the diamond is on the island.

2 · According to this version, it does not follow from the native's two statements that the reasoner will believe the propositions (1) and (2) of the solution to the last problem, but it does follow that he will believe the weaker propositions (1)′ and (2)′ (see the note below). But, as we have seen in the solution of the last problem, this is enough to guarantee that the diamond is on the island.

Note: If a native of a knight-knave island says: "If I am a knave, then X," it logically follows that if X is true, the native must be a knight (because if X is true, then *any* proposition implies X, hence it is true that if the native is a knave, then X, but a knave couldn't make such a true statement). This is why the reasoner (who believes the rules of the island) will believe (1)′. As for (2)′, it is obvious that if a native says, "I am a knight and X," it follows that if the native is a knight, then X must be true.

3 · *Step 1:* The reasoner believes k≡(Bk⊃~D)—by virtue of the native's first statement. Then by Lemma 1 of the last chapter, the reasoner will believe Bk⊃B~D. And so the reasoner reasons: "If I ever believe that he is a knight, then I will believe that the diamond is not on the island. If I should also believe that the diamond *is* on

138

the island, then I will be inconsistent. Therefore, if I should ever believe he is a knight, then his second statement is true. And, of course, if his second statement is true, then he is a knight. This proves that if I should ever believe that he is a knight, then he really is a knight."

Step 2: The reasoner continues: "So suppose I believe he's a knight. Then he really is a knight, as I have just shown, hence his first statement Bk⊃∼D is true. Also, if I believe he's a knight, then Bk is true, and thus ∼D is true. Therefore, if I believe he's a knight, then the diamond is not on the island. He said just this in his first statement, and so he is a knight."

Step 3: The reasoner continues: "Now I believe he is a knight and I have already shown that Bk⊃∼D, hence ∼D is true—the diamond is not on this island."

Step 4: The reasoner now believes that the diamond is not on the island. Therefore, if he should ever believe that the diamond *is* on the island, he will be inconsistent. This proves that the native's second statement was true and therefore the native is in fact a knight. And so the native's first statement was also true—Bk⊃∼D is true. But Bk is true (as we have proved), and so ∼D is true. Therefore the diamond is not on the island.

4 · *Step 1:* The native asserted Bk⊃BD, and so the reasoner believes k≡(Bk⊃BD). Then by Lemma 1 of the last chapter (taking BD for p), the reasoner will believe Bk⊃BBD, and so the reasoner reasons: "Suppose I ever believe that the native is a knight. Then I will believe BD. Then I will believe k and I will believe BD, hence I will believe k&BD. I also believe the Sage's statement (k&BD)⊃D, so if I ever believe k&BD, then I will believe D. Therefore, if I ever believe k, I will also believe D—Bk⊃BD is true. This is what the native said, hence he is a knight."

Step 2: The reasoner continues: "I now believe k—Bk is true. Also Bk⊃BD is true (as I have proved), hence BD is true (I will *believe* the diamond is on the island). Thus k is true and BD is true, so k&BD is true. Then, by the Sage's statement, D must be true—the diamond is on this island."

Step 3: The reasoner now believes D, so BD is true. Then Bk⊃BD is certainly true, hence the native is really a knight. Thus k and BD are both true, so k&BD is true. Hence, by the Sage's statement, D must be true—the diamond is on the island.

5 · The solution to this problem is somewhat simpler than the solution to Problem 4.

Step 1: Since the native said what he did, then by Lemma 3 of the last chapter, the reasoner will believe k⊃BD. Hence he will believe k⊃(k&BD). Since he also believes the Sage's statement (k&BD)⊃D, then he will believe k⊃D. Since the native said that he will believe k⊃D, then the reasoner will believe that the native is a knight—he will believe k. And since he will believe k⊃D, then he will believe D.

Step 2: Since the reasoner will believe k⊃D, then the native is really a knight. Hence k is true and also BD is true (as we have shown). Hence k&BD is true, and since (k&BD)⊃D is true, then D is true. So again the diamond is on the island.

Now we know that the chances are 80 percent that the diamond is on the island, since it is on the island in four out of five equally probable versions. This probability should be high enough to interest the enterprising reader who wishes to search for it.

Discussion. The essential mathematical content of Problem 4 is that for any propositions k and p, if a reasoner of type 4 believes k≡(Bk⊃Bp) and believes (k&Bp)⊃p, then he will believe p. This is a strengthening of Theorem 2 of the last chapter, because if a reasoner of type 4 believes Bp⊃p, he will certainly believe (k&Bp)⊃p, and so the present hypothesis is weaker than that of Theorem 2, Chapter 15 (and we have derived the same conclusion from a weaker assumption).

Likewise, the essential mathematical content of Problem 5 is that if a reasoner of type 4 believes $k\equiv B(k{\supset}p)$ and $(k\&Bp){\supset}p$, then he will believe p. This is stronger than Theorem 3 of the last chapter for the same reasons.

Curiously enough, Theorem 1 of the last chapter does not appear to have an analogous strengthening. If a reasoner of type 4 believes $k\equiv(Bk{\supset}p)$ and $(k\&Bp){\supset}p$, there does not seem to be any reason to conclude that he will believe p.

· 17 ·

Löb's
Island

SOMEWHERE IN the vast reaches of the ocean there is a particularly interesting knight-knave island which I shall refer to as
Löb's Island. Given any person who visits the island, and given any
proposition p, there is at least one native of the island who says to
the visitor: "If you ever believe that I am a knight, then p is
true."

In the problems of this chapter, a reasoner of type 4 visits the
island. It is given that the rules of the island hold (knights tell the
truth, knaves lie, and every native is a knight or a knave) and that
the reasoner believes the rules of the island.

HENKIN'S PROBLEM

Suppose a native of Löb's Island says to the reasoner: "You will
believe that I am a knight." On the surface it seems that there is
no way to tell whether the native is a knight or a knave; it would
appear that maybe the reasoner will believe that the native is a
knight (in which case the native made a true statement, and therefore is a knight), or that maybe the reasoner will never believe that
the native is a knight (in which case the native made a false statement and is accordingly a knave). Is there any way to decide between
these two alternatives?

This problem derives from a famous problem posed by Leon Henkin and answered by M. H. Löb. The surprising thing is that it *is* possible to decide whether the native is a knight or a knave.

1

Under the conditions given above, is the native a knight or a knave? (Solutions are given following Problem 3.)

2

Suppose a native of Löb's Island says to a reasoner of type 4: "You will never believe that I am a knave." Assuming the rules of the island hold (and that the reasoner believes them), is the native a knight or a knave?

3

If a reasoner of type 4 visits Löb's Island (and believes the rules of the island), is it possible for him to be consistent and to believe that he is consistent?

SOLUTIONS TO PROBLEMS 1, 2, AND 3

1 · The solution is a bit tricky. Let P_1 be the native who said: "You will believe I'm a knight." Let k_1 be the proposition that P_1 is a knight. Since the reasoner believes the rules of the island and P_1 said what he did, then the reasoner will believe the proposition $k_1 \equiv Bk_1$. From just this fact, it is *not* possible to determine whether k_1 is true or false; but this island is Löb's Island, hence there is a native P_2 who will say to the reasoner: "If you ever believe I'm a knight, then P_1 is a knight." (Remember that for *any* proposition p, there is some native who says to the reasoner: "If you ever believe I'm a knight, then p.") Let k_2 be the proposition that P_2 is a knight. Since P_2

143

asserted the proposition $Bk_2 \supset k_1$, then the reasoner will believe the proposition $k_2 \equiv (Bk_2 \supset k_1)$. He also believes $Bk_1 \equiv k_1$, hence he believes $Bk_1 \supset k_1$, and so he believes $Bk_1 \supset k_1$ and $k_2 \equiv (Bk_2 \supset k_1)$. Then by Theorem 1, Chapter 15, page 126 (reading "k_2" for "k" and "k_1" for "C"), he will believe k_1. Since P_1 said that the reasoner would believe k_1, then P_1 is a knight. And so P_1 is a knight and the reasoner will believe that P_1 is a knight.

Note: We might remark that even without the assumption that the rules of the island really hold, we can still conclude that the reasoner will *believe* that P_1 is a knight.

2 · To avoid repetition of similar arguments, let us now note once and for all that if a reasoner of type 4 visits Löb's Island, then for any proposition p, if he believes the proposition $Bp \supset p$, he will believe p. (Reason: Since this island is Löb's Island, some native will say to him: "If you ever believe that I'm a knight, then p." Hence the reasoner will believe $k \equiv (Bk \supset p)$, where k is the proposition that the native is a knight. Then, since he believes $Bp \supset p$, he will believe p, by Theorem 1, Chapter 15.)

Now for the problem at hand: The native has told the reasoner, "You will never believe I'm a knave." Then the reasoner believes $k \equiv \sim B \sim k$ (k is the proposition that the native is a knight). Hence he will believe the logically equivalent proposition $\sim k \equiv B \sim k$, hence he will believe $B \sim k \supset \sim k$. Therefore he will believe $Bp \supset p$, when p is the proposition $\sim k$. Then, by the remarks of the last paragraph, he will believe p—i.e., he will believe $\sim k$. And so the reasoner will believe that the native is a knave. Since the native said that the reasoner wouldn't believe that the native is a knave, then the native in fact is a knave.

3 · Suppose a reasoner of type 4 visits Löb's Island. Then for any proposition p, there is some native who will say to him: "If you ever believe I'm a knight, then p." In particular, this is true if we take

for p the proposition $\perp$ (which, we recall, stands for logical falsehood). So there is a native who says to the reasoner: "If you ever believe I'm a knight, then $\perp$." Thus the reasoner believes the proposition $k\equiv(Bk\supset\perp)$. Now, $Bk\supset\perp$ is logically equivalent to $\sim Bk$, hence $k\equiv(Bk\supset\perp)$ is logically equivalent to $k\equiv\sim Bk$, hence the reasoner believes $k\equiv\sim Bk$. Then, according to Theorem 1, Chapter 12, page 101, he cannot believe in his own consistency without becoming inconsistent.

REFLEXIVITY

Reflexive Reasoners. We shall say that a reasoner is *reflexive* if for every proposition q, there is at least one proposition p such that the reasoner will believe $p\equiv(Bp\supset q)$.

Any reasoner who visits Löb's Island (and believes the rules of the island) will automatically become a reflexive reasoner, since for any proposition q, there is at least one native who will say: "If you ever believe I'm a knight, then q is true," and so the reasoner will believe $k\equiv(Bk\supset q)$, where k is the proposition that the native is a knight. However, a reasoner who has never visited Löb's Island might be a reflexive reasoner for completely different reasons (some of which we will consider in Chapter 25).

Let us note that if a *reflexive* reasoner of type 4 visits an *ordinary* knight-knave island (it doesn't have to be Löb's Island) and meets a native who says to him, "You will believe that I'm a knight," then the reasoner *will* believe that the native is a knight. (He doesn't need a second native to tell him, "If I'm a knight, then so is the first native.") Also, if a *reflexive* reasoner of type 4 goes to an *ordinary* knight-knave island and is told by a native, "You will never believe I'm a knave," the reasoner will believe that the native is a knave (as in Problem 2).

Also, no consistent reflexive reasoner of type 4 can believe he is consistent.

And, one more thing: Suppose a *reflexive* reasoner of type 4 is thinking of visiting the knight-knave island with the sulfur baths and mineral waters, and his family doctor, whom he trusts, tells him: "If you believe that the cure will work, then it will." Then without further ado, the reasoner will believe that the cure will work (he doesn't have to go first to the island and meet a native who tells him, "If you believe I'm a knight, then the cure will work").

All these facts are special cases of the following theorem (which springs from Theorem 1 of Chapter 15).

Theorem A (After Löb). For any proposition q, if a reflexive reasoner of type 4 believes Bq⊃q, he will believe q.

Reflexive Systems. Let us now consider the type of mathematical systems described in Chapter 13. We recall that for any system S, for any proposition p expressible in the system, the proposition Bp (p is provable in S) is also expressible in the system. (Remember that for systems, "B" means "provable.") When we have only one system S under discussion, the word "proposition" shall be understood to mean "proposition expressible in S."

We now define S to be *reflexive* if for every proposition q (expressible in S), there is at least one proposition p (expressible in S) such that the proposition p≡(Bp⊃q) is *provable* in S. Theorem A above obviously holds for *systems* as well as reasoners: Given any reflexive system S and any expressible proposition q, if Bq⊃q is provable in S, so is q. This is Löb's Theorem.

We shall call S a *Löbian* system if for every proposition p, if Bp⊃p is provable in the system, so is p. We have now established Löb's Theorem.

Theorem L (Löb's Theorem). Every reflexive system of type 4 is Löbian—i.e., for any reflexive system S of type 4, if Bp⊃p is provable in S, so is p.

Corollary. For any reflexive system of type 4, if p≡Bp is provable in the system, so is p.

Discussion. Gödel proved his incompleteness theorems for several systems, including the system of Arithmetic, which we have briefly mentioned. These systems are all reflexive systems of type 4, and this is what enabled Gödel's arguments to go through. Gödel constructed a sentence g that asserted its own nonprovability in the system (the sentence g≡~Bg is provable in the system).

Later the logician Leon Henkin constructed a sentence h such that h≡Bh is provable in the system, and raised the problem whether there was any way to tell whether h was provable in the system or not. Such a sentence h can be thought of as asserting: "I *am* provable in the system." (It resembles a native who says: "You *will* believe that I'm a knight.") On the surface, it would appear equally possible that h is true and provable in the system, or false and not provable in the system. The problem remained open for several years, and was finally solved by Löb. We have the answer in the corollary above: If the system is reflexive and of type 4, then Henkin's sentence h *is* provable in the system.

Reflexive and Gödelian Systems. We recall that a system is called *Gödelian* if there is some proposition p such that p≡~Bp is provable in the system.

4

Every reflexive system of type 1 is also Gödelian. Why is this? (Solutions are given following Problem 5.)

Strong Reflexivity. We will say that a system S is *strongly* reflexive if for every proposition q, there is a proposition p such that p≡B(p⊃q) is provable in S.

The connections between reflexivity and strong reflexivity are given in the following theorem.

Theorem R (Reflexivity Theorem). It consists of two parts:
(a) Any strongly reflexive system of type 1 is reflexive.
(b) Any reflexive system of type 1* is strongly reflexive.

5

Prove Theorem R.

SOLUTIONS TO PROBLEMS 4 AND 5

4 · Suppose S is reflexive and of type 1. Then for any proposition q, there is some p such that $p \equiv (Bp \supset q)$ is provable in S. We take for q the proposition $\perp$, and so there is some p such that $p \equiv (Bp \supset \perp)$ is provable in S. But $Bp \supset \perp$ is logically equivalent to $\sim Bp$, hence $p \equiv (Bp \supset \perp)$ is logically equivalent to $p \equiv \sim Bp$, and so $p \equiv \sim Bp$ is provable in S, and therefore S is Gödelian. (Actually, since we are basing propositional logic on $\supset$ and $\perp$, the proposition $\sim Bp$ *is* the proposition $Bp \supset \perp$, and so we really didn't even have to assume that S is of type 1.)

5 · Suppose S is of type 1. Let q be any proposition.
(a) Suppose S is strongly reflexive. Then there is a proposition p such that $p \equiv B(p \supset q)$ is provable in S. Since S is of type 1, it follows that $(p \supset q) \equiv (B(p \supset q) \supset q)$ is provable in S. (For any propositions X, Y, and Z, the proposition $(X \supset Z) \equiv (Y \supset Z)$ is a logical consequence of $X \equiv Y$. Taking p for X, $B(p \supset q)$ for Y, and q for Z, the proposition $(p \supset q) \equiv (B(p \supset q) \supset q)$ is a logical consequence of $p \equiv B(p \supset q)$.) Therefore there is a proposition p'—namely, $p \supset q$—such that $p' \equiv (Bp' \supset q)$ is provable in S. Hence S is reflexive.
(b) Suppose also that S is regular (and hence of type 1*) and that S is reflexive. Then there is a proposition p such that $p \equiv (Bp \supset q)$ is

148

provable in S. Since S is regular, it follows that $Bp \equiv B(Bp \supset q)$ is provable, hence $(Bp \supset q) \equiv (B(Bp \supset q) \supset q)$ is provable. We now take p' to be the proposition $Bp \supset q$, and we see that $p' \equiv (Bp' \supset q)$ is provable. Since there is a proposition p' such that $p' \equiv (Bp' \supset q)$ is provable in S, S is strongly reflexive.

Remarks. It of course follows from the above theorem and Löb's Theorem that any strongly reflexive system of type 4 is Löbian. We proved this another way in Theorem 3, Chapter 15.

Part VII

IN DEEPER
WATERS

Reasoners of Type G

MODEST REASONERS

We have called a reasoner *conceited* if for every proposition p, he believes Bp⊃p. Now, if a reasoner believes p, then there is nothing immodest about his believing Bp⊃p. (Indeed, if he believes p and is of type 1, he will also believe q⊃p for *any* proposition q whatsoever, since q⊃p is a logical consequence of p. In particular, he will believe Bp⊃p.)

We shall now call a reasoner *modest* if for every proposition p, he believes Bp⊃p only if he believes p (in other words, if he believes Bp⊃p, then he believes p). In analogy with *systems,* we might also call a modest reasoner a *Löbian* reasoner.

Löb's Theorem states that every reflexive system of type 4 is Löbian. Stated in terms of reasoners, every reflexive reasoner of type 4 is modest.

Many results that can be proved for a given reasoner under the assumption that he is a reflexive reasoner of type 4 can be proved more swiftly from the assumption that he is a modest reasoner of type 4. For example, suppose a modest reasoner of type 4 (or even a modest reasoner of type 1) believes that he is consistent. Then he believes ~B⊥. Hence he believes the logically equivalent proposition B⊥⊃⊥. Then, being modest, he must believe ⊥, which means that he is inconsistent! And so no modest reasoner—even of type 1—can *consistently* believe in his own consistency. (This, of course, doesn't mean that he necessarily *is* inconsistent; he might happen to be consistent, but if he is consistent—and a modest reasoner of type 1 —then he cannot believe he is consistent.)

REASONERS OF TYPE G

Let us say that a reasoner is modest with respect to a given proposition p if it is the case that if he believes Bp⊃p, then he also believes p. A reasoner, then, is modest if he is modest with respect to every proposition p. We now say that a reasoner *believes* he is modest with respect to p if he believes the proposition B(Bp⊃p)⊃Bp—he believes that if he should ever believe that his belief in p implies p, then he will believe p. We now say that a reasoner believes he is modest if for every proposition p, he believes he is modest with respect to p —in other words, for every proposition p, he believes B(Bp⊃p)⊃Bp.

By a *reasoner* of type G is meant a reasoner of type 4 who believes he is modest (for every proposition p, he believes B(Bp⊃p)⊃Bp). By a *system* of type G is meant a system of type 4 such that for every proposition p, the proposition B(Bp⊃p)⊃Bp is provable in the system.

A great deal of research has been going on in recent years about systems of type G. George Boolos has devoted an excellent book† to the subject, which we strongly recommend as a follow-up to this volume.

Several questions about reasoners of type G readily present themselves. If a reasoner of type G *believes* he is modest, is he necessarily modest? We will show shortly that the answer is yes; indeed, we will see that any reasoner of type 1* who believes he is modest must be modest. Now, what about a reasoner of type 4 who *is* modest. Does he necessarily believe he is modest? We will show that the answer

†*The Unprovability of Consistency* (Cambridge University Press, 1979).

is yes, and hence that any reasoner of type 4 is modest if and only if he believes he is modest—in other words if and only if he is of type G. Then we will show a surprising result discovered independently by the logicians Saul Kripke, D. H. J. de Jongh, and Giovanni Sambin—namely, that any reasoner of type 3 who believes he is modest must be of type 4 (and hence of type G).

Another question: We know that any reflexive reasoner of type 4 is modest (this is Löb's Theorem). Is a modest reasoner of type 4 necessarily reflexive? That is, given a modest reasoner of type 4, is it necessarily true that for any proposition q, there is some proposition p such that he believes $p \equiv (Bp \supset q)$? It is not difficult to prove that the answer is yes; we will do this in the next chapter. And so by the end of the next chapter we will have proved that for any reasoner, the following four conditions are equivalent:

(1) He is a modest reasoner of type 4.
(2) He is of type G (he is of type 4 and believes he is modest).
(3) He is of type 3 and believes he is modest.
(4) He is a reflexive reasoner of type 4.

MODESTY AND BELIEF IN ONE'S MODESTY

For any proposition p, let Mp be the proposition $B(Bp \supset p) \supset Bp$. Thus Mp is the proposition that the reasoner is modest with respect to p. We are about to show that any reasoner of type 4 who believes he is modest really is modest. More specifically, we will show the stronger result that for any proposition p, if he believes that he is modest with respect to p (without necessarily believing that he is modest with respect to any other propositions), then he really is modest with respect to p (in fact this holds even for normal reasoners of type 1). The converse of this stronger fact is not necessarily true

—i.e., if a reasoner of type 4 is modest with respect to p, then he does not necessarily believe that he is modest with respect to p. However, we will show that if a reasoner of type 4 is modest with respect to the proposition Mp, then he will believe that he is modest with respect to p (in other words, if he is modest with respect to Mp, then he will believe Mp). From this it will of course follow that if a reasoner of type 4 is modest (with respect to all propositions), then he will know that he is modest.

1

Why is it true that any normal reasoner of type 1 (and hence any reasoner of type 4) who believes he is modest with respect to p must really be modest with respect to p?

2

By virtue of the last problem, given any proposition p, the proposition $B(Mp) \supset Mp$ is *true* for any reasoner of type 4. Prove that any reasoner of type 4 *believes* the proposition $B(Mp) \supset Mp$. (He knows that if he should *believe* that he is modest with respect to p, then he really is modest with respect to p.)

3

Using the last problem, show that any reasoner of type 4 who is modest with respect to Mp will believe that he is modest with respect to p.

It of course follows from Problem 1 that any reasoner of type 4 who believes he is modest, really is modest. And it follows from Problem 3 that if a reasoner of type 4 is modest (with respect to every proposition), then he must believe he is modest (because for every proposition p, he is modest with respect to Mp, hence according to

156

Problem 3, he believes he is modest with respect to p). We have thus established Theorem M.

Theorem M. A reasoner of type 4 is modest if and only if he believes he is modest.

By virtue of Theorem M, a reasoner is of type G if and only if he is a modest reasoner of type 4. Stated in terms of systems rather than reasoners, we have Theorem M_1.

Theorem M_1. For a system of type 4, the following two conditions are equivalent:

(1) For any proposition p, if $Bp{\supset}p$ is provable in the system, so is p.

(2) For any proposition p, the proposition $B(Bp{\supset}p){\supset}Bp$ is provable in the system.

Stated more briefly, a system of type 4 is Löbian if and only if it is of type G.

We proved in the last chapter that every reflexive system of type 4 is Löbian—this is Löb's Theorem. Combining this with Theorem M_1, we have the important Theorem M_2.

Theorem M_2. Every reflexive system of type 4 is of type G.

Another proof of Theorem M_2 will be given in the next chapter.

THE KRIPKE, DE JONGH, SAMBIN THEOREM

We now wish to prove that any reasoner of type 3 who believes that he is modest must also believe that he is normal—and hence must be of type 4, and therefore of type G.

We will actually prove more. Let us say that a reasoner is normal with respect to a proposition p if his believing p implies that he will also believe Bp. A reasoner, then, is normal if and only if he is normal

157

with respect to every proposition p. We will say that a reasoner believes that he is normal with respect to p if he believes the proposition Bp⊃BBp. (Of course if the reasoner *is* normal with respect to p, then the proposition Bp⊃BBp is true.) We will say that a reasoner believes he is normal if for every proposition p, he believes that he is normal with respect to p. A reasoner of type 4, then, is a reasoner of type 3 who believes he is normal. The Kripke, de Jongh, Sambin Theorem states that every reasoner of type 3 who believes he is modest will also believe he is normal (and hence will be of type G). We will prove the stronger result that every reasoner of type 1* who believes he is modest will also believe he is normal. But first we will prove the more elementary result that every reasoner of type 1* who *is* modest is also normal (this result may be new). Then we will prove sharper versions of both these results.

We recall from Chapter 10 that we are using the notation Cp for (p&BP), where we read Cp as "the reasoner *correctly* believes p." The propositions Cp⊃p, Cp⊃Bp, p⊃(Bp≡Cp) are, of course, all tautologies. We first need the following lemma:

Lemma 1. Any reasoner of type 1* believes the following propositions:

(a) BCp⊃BBp
(b) p⊃(BCp⊃Cp)

4

Why is Lemma 1 true?

5

Now show that any modest reasoner of type 1* must be normal. More specifically (and this is a hint!), show that for any proposition p, if a reasoner of type 1* is modest with respect to Cp, then he must be normal with respect to p.

6

Now show that any reasoner of type 1* who believes he is modest must also believe that he is normal. More specifically, show that for any proposition p, if a reasoner of type 1* believes that he is modest with respect to Cp, then he will believe that he is normal with respect to p.

Since every reasoner of type 3 is also of type 1* (as we proved in Chapter 11), we have now established Theorem M₃.

Theorem M₃ (Kripke, de Jongh, Sambin). Every reasoner of type 3 who believes he is modest is of type G.

Stated for systems rather than reasoners, Theorem M₃ states that for any system of type 3, if all propositions of the form B(Bp⊃p)⊃Bp are provable in the system, then all propositions of the form Bp⊃BBp are provable in the system, and hence the system must be of type G.

The following two exercises provide some curious alternative ways of characterizing reasoners of type G.

Exercise 1. Show that for any reasoner of type 4, the following two conditions are equivalent.

(a) The reasoner is of type G.

(b) And, for any propositions p and q, the reasoner believes B(q⊃p)⊃(B(Bp⊃q)⊃Bp).

Exercise 2. Prove that for any reasoner of type 4, the following two conditions are equivalent.

(a) He is of type G.

(b) For any propositions p and q, if he believes (Bp&Bq)⊃p, then he will believe Bq⊃p.

SOLUTIONS

1 · By hypothesis, the reasoner believes Mp—the proposition B(Bp⊃p)⊃Bp. We are to show that he is modest with respect to p —i.e., that if he believes Bp⊃p, then he believes p. So we assume he believes Bp, and we are to show that he believes (or will believe) p.

Since he believes Bp⊃p (by assumption), he believes B(Bp⊃p) (since he is normal). Also, he believes B(Bp⊃p)⊃Bp (by hypothesis). Then, being of type 1, he believes, or will believe, Bp. He also believes Bp⊃p. Hence he believes, or will believe, p.

2 · Essentially, this follows from Problem 1 and the fact that a reasoner of type 4 "knows" that he is of type 4, hence knows how he can reason. In more detail, the reasoner reasons thus:

"Suppose I ever believe Mp—that is, suppose I ever believe B(Bp⊃p)⊃Bp. I am to show that I am modest with respect to p— i.e., that if I ever believe Bp⊃p, then I will believe p. So suppose I believe Bp⊃p; it remains to show that I will believe p.

"And so I will assume that I will believe B(Bp⊃p)⊃Bp and that I will believe Bp⊃p. I must then show that I will believe p. Well, since I will believe Bp⊃p (by my second assumption), then I will believe B(Bp⊃p). Since I also believe B(Bp⊃p)⊃Bp (by my first assumption), then I will believe Bp. Once I believe Bp and Bp⊃p, then I will believe p."

At this point the reasoner has derived Bp from the two assumptions B(Mp) and B(Bp⊃p), he will believe (B(Mp)&B(Bp⊃p))⊃Bp, and the logically equivalent proposition B(MP)⊃(B(Bp⊃p)⊃Bp), which is the proposition B(Mp)⊃Mp.

3 · If a reasoner believes B(Mp)⊃Mp and is modest with respect to Mp, then he will of course believe Mp (because for any proposition q, if he believes Bq⊃q and is modest with respect to q, he will believe

160

q—and so this is the case if q is the proposition Mp). Now a reasoner of type 4 *does* believe B(Mp)⊃Mp (as we showed in the last problem), and so if he is modest with respect to Mp, he will believe Mp (which means he will believe he is modest with respect to p).

4 · The reasoner is assumed to be of type 1*.

(a) Since he is of type 1, he believes the tautology Cp⊃Bp. Being regular, he will then believe BCp⊃BBp.

(b) Since he is of type 1, he believes the tautology Cp⊃p. Being regular, he then believes BCp⊃Bp. Being of type 1, he then believes (p&BCp)⊃(p&Bp), which is a logical consequence of the last proposition. Thus he believes (p&BCp)⊃Cp (because Cp is the proposition p&Bp) and hence he will believe the logically equivalent proposition p⊃(BCp⊃Cp).

5 · Suppose a reasoner of type 1* is modest with respect to Cp. We are to show that he is then normal with respect to p.

Suppose he believes p. He also believes p⊃(BCp⊃Cp), according to (b) of Lemma 1, hence he will believe BCp⊃Cp. Believing this and being modest with respect to Cp, he will believe Cp. He also believes the tautology Cp⊃Bp, and since he believes Cp, he will believe Bp.

This proves that if he believes p, he will believe Bp, hence he is normal with respect to p.

6 · Suppose a reasoner of type 1* believes he is modest with respect to Cp. By (b) of Lemma 1, he believes p⊃(BCp⊃Cp). Since he is regular, he then believes Bp⊃B(BCp⊃Cp). But he also believes B(BCp⊃Cp)⊃BCp, since he believes he is modest with respect to Cp. Believing these last two propositions, he will believe Bp⊃BCp. He also believes BCp⊃BBp, according to (a) of Lemma 1, and will therefore believe Bp⊃BBp—he will believe that he is normal with respect to p.

Solution to Exercise 1. (a) A reasoner of type G will reason: "Suppose B(q⊃p) and B(Bp⊃q). This means I'll believe q⊃p and I'll believe Bp⊃q, hence I'll believe Bp⊃p, and since I am modest, I'll believe p. Thus (B(q⊃p)&B(Bp⊃q))⊃Bp, or, what is logically equivalent, B(q⊃p)⊃(B(Bp⊃q)⊃Bp)."

(b) Suppose that for any proposition p and q, a reasoner of type 4 believes B(q⊃p)⊃(B(Bp⊃q)⊃Bp). Then this is also true if q and p are the same proposition, and so he believes B(p⊃p)⊃(B(Bp⊃p)⊃p). He also believes B(p⊃p)—because he believes the tautology p⊃p, and being normal, he then believes B(p⊃p)—and hence he believes B(Bp⊃p)⊃Bp. Therefore the reasoner is of type G.

Solution to Exercise 2. (a) Suppose a reasoner of type 4 believes (Bp&Bq)⊃p. Then he will reason: "(Bp&Bq)⊃p. Hence Bq⊃(Bp⊃p). I now believe Bq⊃(Bp⊃p). Now, suppose Bq. Then I'll believe Bq, and since I believe Bq⊃(Bp⊃p), I'll believe Bp⊃p. And therefore Bq⊃B(Bp⊃p). Also B(Bp⊃p)⊃Bp, and so Bq⊃Bp. Thus Bq⊃(Bp&Bq). And since (Bp&Bq)⊃p, then Bq⊃p."

(b) Suppose a reasoner of type 4 is such that for any proposition p and q, if he believes (Bp&Bq)⊃p, then he believes Bq⊃p. We will show that he is modest (and hence of type G).

Suppose he believes Bp⊃p. Then for any proposition q, he certainly believes (Bp&Bq)⊃p. Hence, for any proposition q, he believes Bq⊃p. Well, take any proposition q such that he believes q (for example, take q = T). Then he believes Bq, and believing Bq⊃p, he will believe p.

· 19 ·

Modesty, Reflexivity, and Stability

MORE ON REASONERS OF TYPE G

1

There is something very interesting about a consistent reasoner of type G—or even a consistent modest reasoner of type 1*—namely, that there is *no* proposition p such that he can believe that he doesn't believe p! (He cannot believe ~Bp!) Why is this?

2

It hence follows that if a reasoner of type G believes that he doesn't believe p, then he will be inconsistent (even though it may be true that he doesn't believe p). Thus for any reasoner of type G, the proposition B~Bp⊃B⊥ is *true*.

Prove that for any reasoner of type G and any proposition p, the proposition B~Bp⊃B⊥ is not only true, but is *known* to be true by the reasoner.

163

3

This problem is a sharpening of an earlier theorem. Let us recall how we proved that every reflexive reasoner of type 4 is of type G. We did this in two stages: We first proved that every reflexive reasoner of type 4 is Löbian (Löb's Theorem), and then we proved that every Löbian reasoner of type 4 is of type G.

Now, let us consider a reasoner of type 4 who is *not* necessarily reflexive. It might be that for *some* proposition q, there is a proposition p such that the reasoner believes p≡(Bp⊃q), and for some other proposition q, there is no such proposition p. This much we do know: If, for a given q, there is some p such that the reasoner believes p≡(Bp⊃q), then if the reasoner believes Bq⊃q, he will also believe q (by Theorem 1, Chapter 15), and hence the proposition B(Bq⊃q)⊃Bq is *true*. But does that mean that the reasoner necessarily *knows* that it is true? The answer is the solution to this problem:

Prove that for any propositions p and q, if a reasoner of type 4 believes p≡(Bp⊃q), then he *believes* B(Bq⊃q)⊃Bq.

Discussion. Of course the solution to Problem 3 yields an alternative proof that any reflexive reasoner of type 4 must be of type G. I will now mention something else worth noting.

By virtue of Problem 3, given any propositions p and q, the proposition B(p≡(Bp⊃q))⊃(B(Bq⊃q)⊃Bq) is *true* for any reasoner of type 4. It can be shown that any reasoner of type 4 *knows* that the above proposition is true. The reader might try this as an exercise.

SOME FIXED-POINT PRINCIPLES[†]

We have now seen two different proofs that every reflexive reasoner of type 4 is of type G. We will shortly prove that every reasoner of

[†]These are special cases of a remarkable fact discussed in the final chapter.

type G is reflexive—for every q, there is some p such that he believes p≡(Bp⊃q). But first some preliminary problems.

4

I have already warned you of the dangers of believing any proposition of the form p≡~Bp. If, however, you happen to be a reasoner of type G, then I'm afraid you have no alternative!

Given a reasoner of type G, find a proposition p such that he must believe p≡~Bp.

5

It is also true that given any reasoner of type G, there is a proposition p such that he believes p≡B~p. Prove this.

6

Given a proposition q, find a proposition p (expressible in terms of q) such that any reasoner of type G will believe p≡B(p⊃q).

7

Do the same with Bp⊃q—i.e., find a proposition p such that any reasoner of type G will believe p≡(Bp⊃q).

Note: The result of the last problem is that every reasoner of type G is reflexive. We have already proved that every reflexive reasoner of type 4 is of type G, and so we now have Theorem L*.

Theorem L*. A reasoner of type 4 is of type G if and only if he is reflexive.

SOME MORE FIXED-POINT PROPERTIES

8

Given a reasoner of type G and any propositions p and q, show that if the reasoner believes $p \equiv B(p \supset q)$, then he will believe $p \equiv Bq$.

9

Show that if a reasoner of type G believes $p \equiv (Bp \supset q)$, then he will believe $p \equiv (Bq \supset q)$.

10

A reasoner of type G goes to a knight-knave island (and believes the rules of the island), and asks a native whether he is married. The native replies: "You will believe that either I am a knight or I am married."

Will the reasoner necessarily believe that the native is married? Will he necessarily believe that he is a knight?

STABILITY

In preparation for the next chapter, we will now introduce the notion of *stability*.

We will call a reasoner *stable* if for every proposition p, if he believes that he believes p, then he really does believe p. We will call a reasoner *unstable* if he is not stable—i.e., if there is at least one proposition p such that the reasoner believes that he believes p, but he does not actually believe p. Of course every accurate reasoner is automatically stable, but stability is a much weaker condition than accuracy. An unstable reasoner is inaccurate in a very strange way; indeed, instability is as strange a psychological characteristic as peculiarity.

We note that stability is the converse of normality. If a *normal* reasoner believes p, then he believes Bp, whereas if a *stable* reasoner believes Bp, then he believes p.

We shall use the terms *stable* and *unstable* for mathematical systems as well as reasoners. We will call a mathematical system S *stable* if for any proposition p, if Bp is provable in S, then so is p. We shall say that a reasoner is stable with respect to a particular proposition p if the proposition BBp⊃Bp is true—i.e., if his belief that he believes p guarantees that he really does believe p. We shall say that he *believes* he is stable with respect to p if he believes the proposition BBp⊃Bp. Finally, we will say that he believes that he is stable if for every proposition p, he believes BBp⊃Bp (for every p, he believes that he is stable with respect to p).

11

Prove that if a modest reasoner believes that he is stable, then he is either unstable or inconsistent.

Remark. The above result of course implies that no consistent stable reasoner of type G can ever know that he is stable.

SOLUTIONS

1 · Suppose a modest reasoner of type 1* believes ~Bp. He also believes the tautology ⊥⊃p, hence he believes B⊥⊃Bp (because he is regular), hence he believes the logically equivalent proposition ~Bp⊃~B⊥, hence he believes ~Bp⊃(B⊥⊃⊥). Since he believes ~Bp,

he then believes $(B\bot\supset\bot)$. Then, since he is modest, he will believe $\bot$, which means that he will be inconsistent. Therefore, if he is consistent (and a modest reasoner of type 1*), he will never believe $\sim Bp$.

2 · Any reasoner of type 4 (or even any normal reasoner of type 1*) will successively believe the following propositions:

(1) $\bot\supset p$
(2) $B\bot\supset Bp$
(3) $\sim Bp\supset\sim B\bot$
(4) $\sim B\bot\supset(B\bot\supset\bot)$—this is a tautology
(5) $\sim Bp\supset(B\bot\supset\bot)$—by (3) and (4)
(6) $B\sim Bp\supset B(B\bot\supset\bot)$

If the reasoner is of type G, he will also believe $B(B\bot\supset\bot)\supset B\bot$, hence he will believe $B\sim Bp\supset B\bot$.

3 · Suppose a reasoner of type 4 believes $p\equiv(Bp\supset q)$. We showed in Lemma 1, Chapter 15, page 125, that he will then believe $Bp\supset Bq$. Since he is normal, he will believe that he believes $p\equiv(Bp\supset q)$ and he will believe that he believes $Bp\supset Bq$. And so he reasons: "I believe $p\equiv(Bp\supset q)$, and I believe $Bp\supset Bq$. Now, suppose I ever believe $Bq\supset q$. Then, since I believe $Bp\supset Bq$, I will believe $Bp\supset q$. And, since I believe $p\equiv(Bp\supset q)$, I will believe p. Then I will believe Bp, and since I believe $Bp\supset q$, I will believe q. This shows that if I ever believe $Bq\supset q$, I will believe q."

At this point the reasoner believes $B(Bq\supset q)\supset Bq$.

4 and 5 · We will first solve Problem 5. Take p to be $B\bot$. We claim that any reasoner of type G believes $B\bot\equiv B\sim B\bot$ (and hence believes $p\equiv B\sim p$, where p is the proposition $B\bot$).

We showed in Problem 2 that for *any* proposition p, a reasoner of type G believes $B\sim Bp\supset B\bot$, hence (taking $\bot$ for p) he believes $B\sim B\bot\supset B\bot$. Also, he believes the tautology $\bot\supset\sim B\bot$, so he believes

B⊃B∼B⊥. And since he believes B∼B⊥⊃B⊥, he must believe B⊥≡B∼B⊥.

Now for the solution of Problem 4. We have just shown that the reasoner believes B⊥≡B∼B⊥, hence he believes ∼B⊥≡∼B∼B⊥. And so he believes p≡∼Bp, when p is now the proposition ∼B⊥.

Translated into words, a reasoner of type G believes the proposition that he is consistent if and only if he doesn't believe he is consistent. He also believes the proposition that he is inconsistent if and only if he believes that he is consistent.

6 · A solution is to take p to be Bq. Let us verify that this works. The reasoner believes the tautology q⊃(Bq⊃q), and since he is regular, he then believes Bq⊃B(Bq⊃q). He also believes B(Bq⊃q)⊃Bq (since he is of type G), hence he must believe Bq≡B(Bq⊃q). Therefore he believes p≡B(p⊃q), where p is the proposition Bq.

7 · We have shown that the reasoner believes Bq≡B(Bq⊃q). Then, by propositional logic, he will believe (Bq⊃q)≡B(Bq⊃q)⊃q. And so he will believe p≡(Bp⊃q), where p is now the proposition Bq⊃q. (Note: We have just duplicated the proof of Theorem R, Chapter 17, page 148. According to Problem 6, the reasoner is strongly reflexive, and so by (a) of Theorem R, he is reflexive.)

8 · Suppose he believes p≡B(p⊃q). Then by Lemma 3, Chapter 15, page 129, he will believe p⊃Bq. He also believes the tautology q⊃(p⊃q), and being regular, he then believes Bq⊃B(p⊃q). Also, since he believes p≡B(p⊃q), he must believe B(p⊃q)⊃p. Since he believes Bq⊃B(p⊃q) and B(p⊃q)⊃p, he will believe Bq⊃p. And so he believes Bq⊃p and p⊃Bq (as we have shown), hence he must believe p≡Bq.

9 · Suppose he believes p≡(Bp⊃q). Then he will believe p⊃(Bp⊃q), and also, by Lemma 1, Chapter 15, he will believe Bp⊃Bq. He believes the tautology q⊃(Bp⊃q), hence he will believe q⊃p (since he also believes p≡(Bp⊃q)). But he is regular, hence he will then

169

believe Bq⊃Bp. Believing this, together with Bp⊃Bq, he will believe Bp≡Bq. Then, using propositional logic, he will believe (Bp⊃q)≡(Bq⊃q). Then, since he believes p≡(Bp⊃q), he will believe p≡(Bq⊃q).

10 · The reasoner won't have any idea whether the native is married or not, but he will believe that the native is a knight. We can see this as follows.

Let m be the proposition that the native is married. The native has asserted B(kvm), where k is the proposition that the native is a knight. Hence the reasoner will believe k≡B(kvm). Then he will certainly believe k⊃B(kvm). He also believes the tautology k⊃(kvm), hence he will believe Bk⊃B(kvm), since he is regular. He also believes B(kvm)⊃k, since he believes k≡B(k⊃m). Hence he will believe Bk⊃k. Then, being of type G, he will believe k.

11 · Suppose he believes that he is stable. Then for every proposition p, he believes BBp⊃Bp, hence he believes BB⊥⊃B⊥. If he is modest, he will then believe B⊥ (because for *any* proposition q, a modest reasoner who believes Bq⊃q will believe q, and so this is true in particular if q is the proposition B⊥). Since he believes B⊥, then if he is stable, he will believe ⊥ and thus be inconsistent. This proves that if he believes BB⊥⊃B⊥, he cannot be modest, stable, and consistent. Therefore, if he is modest, stable, and consistent, he can never believe BB⊥⊃B⊥, and so he cannot know that he is stable with respect to ⊥.

· *Part VIII* ·

CAN'T DECIDE!

· 20 ·

Forever
Undecided

WE HAVE discussed Gödel's Second Incompleteness Theorem
(no consistent Gödelian system of type 4 can prove its own consistency), but we have not yet discussed his First Incompleteness Theorem. We now turn to this—first in the context of a reasoner on a
knight-knave island.

We recall from the last chapter that a reasoner is called *stable* if
for every proposition p, if he believes Bp, then he believes p.

AN INCOMPLETENESS
PROBLEM

We shall say that a reasoner's belief system is *incomplete* if there
is at least one proposition p such that the reasoner will never believe
p and never believe ~p (he will be forever undecided as to whether
p is true or false).

The following problem is modeled after Gödel's First Incompleteness Theorem.

1

A normal reasoner of type 1 comes to the Island of Knights and
Knaves and believes the rules of the island. (Whether the rules really

hold or not is immaterial.) He meets a native who says: "You will never believe that I am a knight."

Prove that if the reasoner is both consistent and stable, then his belief system is incomplete. More specifically, find a proposition p such that the following two conditions hold:

(a) If the reasoner is consistent, then he will never believe p.

(b) If the reasoner is both consistent and stable, then he will never believe ~p.

2 · A Dual of 1

Suppose that the native had instead said: "You will believe that I'm a knave." Now find a proposition p such that the conclusions (a) and (b) of Problem 1 hold.

The same reasoning used in the solution of Problem 1, when applied to mathematical systems rather than reasoners, establishes the following form of Gödel's First Incompleteness Theorem.

Theorem 1. Any consistent, normal, stable Gödelian system of type 1 must be incomplete. More specifically, if S is a normal system of type 1 and if p is a proposition such that p≡~Bp is provable in S, then if S is consistent, p is not provable in S, and if S is also stable, then ~p is also not provable in S.

A proposition p is said to be *undecidable* in a system S if neither it nor its negation ~p is provable in S. Thus Gödel's First Incompleteness Theorem tells us that given any consistent, normal, stable Gödelian system S, there must always be at least one proposition p which, though *expressible* in the language of S, is not *decidable* in S—it can neither be proved nor disproved in S.

3 · A Variant of 1

Suppose the native instead says: "You will believe that you will never believe that I'm a knight." Now find a proposition p satisfying conclusions (a) and (b) of Problem 1.

ω·CONSISTENCY

A *natural* number is by definition either 0 or a positive whole number 1, 2, 3, ... We will henceforth use the word "number" to mean "natural number." Now, consider a property P of (natural) numbers. For any number n, we write P(n) to mean that n has the property P. For example, if P is the property of being an even number, then P(n) means that n is an even number, in which case P(0), P(2), P(4), . . . are all true propositions; P(1), P(3), P(5), . . . are all false propositions. On the other hand, if P is the property of being an odd number, then P(0), P(2), P(4) . . . are all false propositions, whereas P(1), P(3), P(5), . . . are true ones.

The standard symbol in logic for "there exists" is the symbol "∃," which is technically known as the *existential quantifier*. For any property P of numbers, the proposition that there exists at least one number n having the property P is written: ∃nP(n). Now, suppose we have a mathematical system and a property P such that the proposition ∃nP(n) is provable in the system, yet for each particular n, the proposition ∼P(n) is provable—that is, all the infinitely many propositions ∼P(0), ∼P(1), ∼P(2), . . ., ∼P(n) . . . are provable. This means that on the one hand, the system can prove the general statement that some number has the property P, yet each *particular* number can be proved *not* to have the property! Something is clearly wrong with the system, because if ∃nP(n) is true, it is impossible that all the propositions ∼P(0), ∼P(1), . . ., ∼P(n), . . . are also true. Yet, such a system is not necessarily inconsistent—one cannot necessarily derive a formal contradiction from all these propositions. There is, however, a name for such systems. They are called *ω-inconsistent*. (The symbol "ω" is sometimes used to mean the set of natural numbers.)

Let us consider the following analogous situation. Suppose someone gives you a check that says, "Payable at some bank." Assuming that there are only finitely many banks in the world, you can in a finite length of time verify whether the check is good or bad; you simply try cashing it at every bank. If at least one bank accepts it, then you know that the check is good; if every bank rejects it, then you have positive proof that the check is bad. But now suppose you are living in a universe in which there are infinitely many banks, each bank being numbered with a natural number. There is Bank 0, Bank 1, Bank 2, . . . , and so forth. Let us also assume that you are immortal, so that you have infinitely many days ahead of you in which to try to cash the check. Now suppose that in fact no bank will ever cash the check. Then the check was in fact a bad one, yet *at no finite time can you prove it!* You might try the first hundred billion banks and they all refuse the check. You can't offer this as evidence that the one who gave you the check is dishonest; he can always say, "Wait, don't call me dishonest; you haven't tried all the banks yet!" And so, you can never get an actual inconsistency; all you have is an ω-inconsistency (and even this you will never know in any finite length of time).

The notion of ω-inconsistency was once humorously characterized by the mathematician Paul Halmos, who defined an ω-inconsistent mother as one who says to her child: "There is something you can do, but you can't do this and you can't do that and you can't do this other thing . . ." The child says: "But, Ma, isn't there *something* I can do?" The mother replies: "Oh yes, but it's not this, nor that, nor . . ."

A system is called ω-*consistent* if it is not ω-inconsistent. Thus for an ω-consistent system, if $\exists nP(n)$ is provable, then there is at least one number n such that the proposition $\sim P(n)$ is *not* provable. An inconsistent system of type 1 is also ω-inconsistent, because all propositions are provable in an inconsistent system of type 1. Stated otherwise, for systems of type 1, ω-consistency automatically implies (ordinary) consistency. When ω-consistency is being discussed, the term *simple consistency* is sometimes used to mean consistency (this in order to prevent any possibility of confusion). And so, in these terms, any ω-consistent system of type 1 is also simply consistent.

We now go back to the study of reasoners. In all the problems we have considered so far, the *order* in which the reasoner has believed various propositions has played no role. In the remaining problem of this chapter, order will play a key role.

The reasoner comes to the Island of Knights and Knaves on a certain day which we will call the 0th day. The next day is called the 1st day; the day after that is called the 2nd day; and so forth. For each natural number n, we have the nth day, and we assume the reasoner to be immortal and to have infinitely many days ahead of him. For every natural number n and any proposition p, we let $B_n p$ be the proposition that the reasoner believes p sometime during the nth day. The proposition Bp is, as usual, the proposition that the reasoner believes p on some day or other, or, what is the same thing, $\exists n B_n p$ (there exists some n such that the reasoner believes p on the nth day). We shall call the reasoner ω-inconsistent if there is at least one proposition p such that the reasoner (sometime or other) believes Bp, yet for each particular n, he (sometime or other) believes $\sim B_n p$. The reasoner will be called *ω-consistent* if he is not ω-inconsistent.

We now consider a reasoner who satisfies the following three conditions.

Condition C_1. He is of type 1.

Condition C_2. For any natural number n and any proposition p: (a) if the reasoner believes p on the nth day, he will (sooner or later) believe $B_n p$; (b) if he doesn't believe p on the nth day, he will (sooner or later) believe $\sim B_n p$. (The idea is that the reasoner keeps track of what propositions he has believed and has not believed on all past days.)

Condition C_3. For any n and any p, the reasoner believes the proposition $B_n p \supset Bp$ (which, of course, is a true proposition).

The following problem comes very close to Gödel's original version of his First Incompleteness Theorem.

4 · (After Gödel)

The reasoner, satisfying the three conditions above, comes to the Island of Knights and Knaves and believes the rules of the island. He meets a native who says to him, "You will never believe that I am a knight." Prove:

(a) If the reasoner is (simply) consistent, then he will never believe that the native is a knight.

(b) If the reasoner is ω-consistent, then he will never believe that the native is a knave.

Thus if the reasoner is ω-consistent (and hence also simply consistent), then he will remain forever undecided as to whether the native is a knight or a knave.

SOLUTIONS

1 · The proposition p in question is simply the proposition k—the proposition that the native is a knight.

The native has asserted $\sim$Bk, hence the reasoner will believe $k\equiv\sim Bk$.

(a) Suppose the reasoner believes k. Then, being normal, he will believe Bk. He will also believe $\sim$Bk (since he believes k and believes $k\equiv\sim Bk$ and he is of type 1), hence he will be inconsistent. Therefore, if he is consistent, he will never believe k.

(b) Since the reasoner is of type 1 and believes $k\equiv\sim Bk$, he also believes $\sim k\equiv Bk$. Now, suppose he ever believes $\sim k$. Then he will believe Bk. If he is stable, he will then believe k and hence become inconsistent (since he believes $\sim k$). Therefore, if he is both stable and consistent, he will never believe $\sim k$.

In summary, if he is both stable and consistent, he will never believe that the native is a knight and he will never believe that the native is a knave.

2 · We see from the above solution that for *any* proposition p (it doesn't have to be the particular proposition k), if a normal reasoner of type 1 believes $p \equiv \sim Bp$, then if he is consistent, he will never believe p, and if he is also stable, he will never believe $\sim p$. Now, in the present problem, the reasoner believes $k \equiv B \sim k$. Therefore (being of type 1) he believes $\sim k \equiv \sim B \sim k$, and so he believes $p \equiv \sim Bp$, where p is the proposition $\sim k$. Therefore, if he is consistent, he will never believe that the native is a knave, and if he is also stable, then he will never believe that the native is a knight.

3 · A proposition p that now works is $\sim Bk$, as we will show.

The reasoner believes $k \equiv B \sim Bk$.

(a) Suppose he believes $\sim Bk$. Then, being normal, he will believe $B \sim Bk$, and will then believe k (since he believes $k \equiv B \sim Bk$ and is of type 1). Believing k and being normal, he will believe Bk. Thus he will believe both Bk and $\sim Bk$, hence he will become inconsistent. Therefore, if he remains consistent, he will never believe Bk.

(b) Suppose he believes $\sim \sim Bk$. Then he will believe Bk. If he is stable, he will then believe k. Next, he will believe $B \sim Bk$ (since he believes $k \equiv B \sim Bk$ and he is of type 1). Then (under the assumption that he is stable), he will believe $\sim Bk$. Thus he will become inconsistent (since he believes $\sim \sim Bk$). This proves that if the reasoner is both consistent and stable, he will never believe $\sim \sim Bk$.

4 · Having solved Problem 1, the easiest way to solve this problem is to show that any reasoner satisfying Conditions 1, 2, and 3 must be normal, and if he is ω-consistent, he must also be stable.

(a) To show he is normal. Suppose he believes p. Then for some n, he believes p on the nth day. Then by (a) at Condition 2, he will believe $B_n p$. He also believes $B_n p \supset Bp$ (by Condition 3), hence being of type 1 (Condition 1), he will then believe Bp. Therefore he is normal.

(b) Now, suppose he is ω-consistent. We will show that he is

179

stable. Suppose he believes Bp. If he never believes p, then for every number n, he fails to believe p on the nth day, and hence by (b) of Condition 2, for every n he will believe $\sim B_n p$. But since he believes Bp, he will then be ω-inconsistent. Therefore, if he is ω-consistent and believes Bp, he must believe p on some day or other. This proves that if he is ω-consistent, he must be stable (assuming he satisfies Conditions 1, 2, 3—or even just (b) of Condition 2).

Therefore, by Problem 1, he will remain forever undecided.

· 21 ·

More
Indecisions!

ROSSER-TYPE REASONERS

Gödel proved a whole family of mathematical systems to be incomplete under the assumption that they were *ω-consistent*. J. Barkley Rosser subsequently discovered an ingenious method of showing these systems to be incomplete under the weaker assumption that they were *simply* consistent. The undecidable sentence constructed by Rosser is more complicated than Gödel's, but its undecidability can be established under the mere assumption of simple consistency.

Let us return to the reasoners on a knight-knave island where the *order* in which the reasoner believes various propositions makes a difference. For any propositions p and q, we will say that the reasoner believes p *before* he believes q if there is some day on which he believes p and has not yet believed q. If the reasoner *never* believes q, but believes p (on some day or other), then we take it as *true* that he believes p before he believes q. (In other words, he doesn't have to ever believe q in order to believe p before he believes q.) We let $Bp < Bq$ be the proposition that the reasoner believes p before he believes q. If $Bp < Bq$ is true, then $Bq < Bp$ is obviously false.

We shall now define a *Rosser-type* reasoner as a reasoner of type 1 such that the following condition holds.

Condition R. For any propositions p and q, if the reasoner believes p on some day on which he has not yet believed q, then he will (sooner or later) believe $Bp < Bq$ and $\sim(Bq < Bp)$.

The idea behind Condition R is that the reasoner has a perfect memory for what he has and has not believed on all past days. If he believes p before he believes q, then on the first day that he believes p, he hasn't believed q yet (and perhaps never will, or then again he may sometime in the future), and so on any subsequent day he will remember that on the first day on which he believed p, he hadn't yet believed q, and so he will believe $Bp < Bq$ and $\sim(Bq < Bp)$.

1

A Rosser-type reasoner comes to the Island of Knights and Knaves and believes the rules of the island. He meets a native who says to him: "You will never believe I'm a knight before you believe I'm a knave." (Rendered symbolically, the native is asserting the proposition $\sim(Bk < B\sim k)$.)

Prove that if the reasoner is *simply* consistent, then he must remain forever undecided as to whether the native is a knight or a knave.

2

Suppose the native instead said: "You *will* believe I'm a knave before you believe I'm a knight." Does the same conclusion follow?

Discussion. The provable propositions of mathematical systems are provable at various stages. We might think of a mathematical system as a computer programmed to prove various propositions *sequentially.* We say that p is provable *before* p (in a given mathematical system) if p is proved at some stage at which q has not yet been proved (q might or might not be proved at some later stage). For any propositions p and q expressible in the system, the proposition Bp < Bq (p is provable before q) is also expressible in systems of the type considered by Gödel, and Rosser showed that if p *is* provable before q, then the proposition Bp < Bq and the proposition ~(Bq < Bp) are both provable in the system. Rosser also found a proposition p such that p≡~(Bp < B~p) is provable in the system. (Such a proposition p corresponds to the native of Problem 1 who says: "You will never believe I'm a knight before you believe I'm a knave.") Then, by the argument of the solution of Problem 1, if p is provable, then the system is inconsistent, and if ~p is provable, then the system is again inconsistent. And so, if the system is consistent, the proposition p is undecidable in the system.

Gödel's sentence can be paraphrased: "I am not provable at any stage." Rosser's more elaborate sentence can be paraphrased: "I cannot be proved at any stage, unless my negation has been proved earlier." Gödel's sentence, though simpler, requires the assumption of ω-consistency to make the argument go through. Rosser's sentence, though more complicated, works under the weaker assumption of simple consistency.

A SIMPLER INCOMPLETENESS PROBLEM

We have now discussed two incompleteness proofs: Gödel's and Rosser's. There is another one simpler than either, which combines Gödel's method with the use of the notion of *truth*—a notion introduced later by the logician Alfred Tarski. It has always been a puzzle to me why this simple proof—so well known to the experts —is so neglected in elementary textbooks.

In the problem that follows, the order in which the reasoner believes various propositions makes absolutely no difference.

3

Suppose we have a reasoner—call him Paul—who is always *accurate* in his beliefs (he never believes any false propositions). He doesn't have to be of type 1, or normal, nor is it necessary that he actually visit the Island of Knights and Knaves. All we need to know about him is that he is accurate.

One day a native says about him: "Paul will never believe that I'm a knight." It then logically follows that Paul's belief system must be incomplete. Why is this?

4

Suppose the native instead says: "Paul will one day believe that I'm a knave." Would it still follow that Paul's belief system is incomplete?

A MORE SERIOUS PREDICAMENT

5

Let us now consider a consistent stable reasoner of type G. There is one very important question about which he must remain forever undecided—namely, the question of his own consistency. He can never decide whether or not he is consistent. Why is this?

A Question. Of course, the above result holds good replacing "reasoner" with "system": A consistent stable system of type G can never prove its own consistency, nor its own inconsistency.

However, an important question arises: How do we know if there *are* any consistent stable systems of type G? Isn't it possible that the very notion of a consistent stable system of type G conceals some subtle contradiction?

This matter will be fully resolved before we come to the end of this book.

SOLUTIONS

1 · Since the native asserted ∼(Bk < B∼k), the reasoner will believe k≡∼(Bk < B∼k). Suppose that the reasoner is (simply) consistent. We are to show that he will never believe k and never believe ∼k.

(a) Suppose he ever believes k. Since he is consistent, he will never believe ∼k, hence he will believe k before he believes ∼k. Hence he will believe Bk < B∼k (by Condition R). But he also believes k≡∼(Bk < B∼k), hence he will believe ∼k, and believing k, he will be inconsistent! So if he is consistent, he can never believe k.

(b) Suppose he ever believes ∼k. Being consistent, he will never believe k, hence he will believe ∼k before he believes k, hence by Condition R he will believe ∼(Bk < B∼k). But he believes k≡∼(Bk < B∼k), so he will then believe k and be inconsistent. And so, if he is consistent, he cannot believe ∼k either.

2 · The answer is yes. We leave the proof to the reader.

3 · If Paul ever believes that the native is a knight, this will falsify what the native said, thus making the native a knave, and hence making Paul inaccurate in believing that the native is a knight. But we are given that Paul is accurate, hence he won't ever believe that the native is a knight. Hence what the native said is true, so the native is in fact a knight. Then, since Paul is accurate, he will never have the wrong belief that the native is a knave. And so Paul will never know whether the native is a knight or a knave.

Discussion. The mathematical content of the above puzzle is this: In the systems investigated by Gödel, we have not only certain propositions called *provable* propositions, but also a larger class of propositions called the *true* propositions of the system. The class of true propositions of the system is faithful to the truth table rules for the logical connectives; also, for any proposition p of the system, the proposition Bp is a *true* proposition of the system if and only if p is a *provable* proposition of the system. Now, Gödel found a remarkable proposition g such that the proposition g≡~Bg was a *true* proposition of the system (in fact, even provable in the system, but this stronger fact is not needed for the present argument). If g were false, Bg would be true, hence g would be provable, hence true, and we would have a contradiction. Therefore g is true, hence ~Bg is true, hence g is not provable in the system. So g is true but not provable in the system. Since g is true, ~g is false, hence also not provable in the system (since all the provable propositions are true). And so g is undecidable in the system.

4 · The answer is yes. We leave the proof to the reader.

5 · We showed in Chapter 18 that every reasoner of type G is modest, and we showed that no consistent modest reasoner of type 4 (or even type 1) can believe that he is consistent. Therefore no consistent reasoner of type G can know that he is consistent.

For the other half, any stable reasoner who believes that he is inconsistent really is inconsistent, because if he believes that he is inconsistent, then he believes B⊥, and if he is also stable, he believes ⊥, and is hence inconsistent.

Therefore, no stable consistent reasoner of type G can ever believe he is consistent or ever believe he is inconsistent. He is doomed to eternal uncertainty on this issue.

· Part IX ·

POSSIBLE
WORLDS

· 22 ·

It Ain't
Necessarily So!

MUCH OF what we have been doing in this book ties in closely
with the field known as modal logic. The amazing thing about this
field is that it arose out of purely philosophical considerations, but
the axiom systems that have come out of it have recently turned out
to have an entirely different interpretation, which is of mathematical
interest and which figures prominently today in proof theory, computer science, and artificial intelligence. We will have more to say
about the mathematical interpretation in later chapters.

The fundamental concept of modal logic is that of a proposition
being *necessarily* true rather than just true as a matter of fact. Many
times we say: "Yes, it's true that it turned out this way, but it didn't
really have to. It *could* have turned out otherwise." At other times
we say: "Oh, it *had* to turn out this way. It couldn't have been
otherwise." And so we often make a distinction between something
just *happening* to be true, and something being *necessarily* true. As
an example, it happens to be a matter of fact that there are exactly
nine planets in our solar system, but it is perfectly conceivable that
things could have been otherwise and that there could have been
more or less than nine planets. On the other hand, a proposition
such as two plus two is four is not only true, but *necessarily* true. In
no possible circumstances could it be true that two plus two is not
four.

Rather than go further at this point into the philosophy of

189

necessary truth, we will turn to some logic puzzles illustrative of what is known as Kripke semantics, which we will discuss in the next chapter. In preparation for this, let us establish our notation.

Following the modal logician C. I. Lewis, we shall use the letter N for *necessarily* true. (The more usual symbol today is □.) Thus for any proposition p, we read Np as "p is *necessarily* true." And so our notation will be like that of the last several chapters, except that we will be using the letter N instead of B. The definition of a set of propositions being of type 1, 2, 3, 4, or G is the same as before, the only difference again being that we use N in place of B.

1 · A Universe of Reasoners

We now consider an entire universe of reasoners—we will call this universe U_1. Given any proposition p, each reasoner either believes p or disbelieves p, but not both. (Disbelieving a proposition means believing it to be false.) Each reasoner disbelieves ⊥ and his beliefs follow the truth table rules for the logical connectives. For example, he believes p⊃q if and only if he either disbelieves p or believes q (or both). It follows from this that each reasoner believes all tautologies. Also, if a reasoner believes p and believes p⊃q, he must believe q (for if he disbelieves q, he would believe p and disbelieve q, hence would disbelieve p⊃q instead of believing p⊃q). Therefore each reasoner is of type 1. We are also given that each reasoner knows what every other reasoner believes.

Now comes the curious thing. For some reason or other, each reasoner has complete confidence in the judgment of his or her parents, and so for any proposition p, a reasoner believes that p is *necessarily* true if and only if his or her *parents* both believe p! This is known as the "fundamental rule" of the universe. It is so important that we will formally record it.

Fundamental Rule of U_1. A reasoner believes Np if and only if his parents both believe p.

Remarks. There is a rumor that a song writer from our universe— an American composer, in fact—upon visiting the universe U_1 and hearing the reason why the inhabitants believed a proposition to be necessarily true, shook his head skeptically, and said: "It ain't necessarily so!" But I can't vouch for this—I heard it only as a rumor.

A proposition p is called established (for the universe U_1) if all the inhabitants believe it.

Obviously all tautologies are established, but the set of established propositions goes beyond tautologies. In fact, the set of established propositions must be of type 3—that is:

(1a) All tautologies are established.

(1b) If p and p⊃q are established, so is q.

(2) (Np&N(p⊃q))⊃Nq is established.

(3) If p is established, so is Np.

Prove that the set of established propositions is of type 3.

2 · A Second Universe

We now visit another universe, which we will call U_2. The conditions defining this universe are like those of U_1, with one important difference. In this universe, a reasoner believes that p is necessarily true if and only if all of his *ancestors* believe p. (To simplify matters, we shall assume that all the inhabitants are immortal, and so all the ancestors of any person are still living.)

Let us record this fundamental fact.

Fact 2. In the universe U_2, a reasoner x believes Np if and only if all ancestors of x believe p.

191

Now things get more interesting.

Prove that the set of established propositions of the universe U_2 must be of type 4.

3 · A Third Universe

So far, we have left open the question of whether or not the universe had a beginning in time. Well, we now consider a third universe U_3 satisfying all the conditions that we have given for U_2, plus the condition that it *did* have a beginning in time. This means that for each individual x, if we take an ancestor x′ of x, then an ancestor x″ of x′, and keep going backwards in this manner, we must sooner or later come to an ancestor who himself has no ancestors—and hence no parents. (To answer the question of how these parentless individuals came into existence goes beyond the scope of this book. The interested reader should consult either a book on evolution or a book on creationism, depending on his or her scientific or theological interests.)

As the reader may have anticipated, we aim to show that the established propositions of this universe form a class of type G. But before that, we must clarify a point for the reader unfamiliar with the logic of *all* and *some*.

Suppose someone says about a certain club, "All Frenchmen in this club wear berets," and it turns out that there are no Frenchmen in the club. Should the statement then be regarded as false, true, or inapplicable? Those unfamiliar with formal logic may well have different opinions on the matter, but the convention adopted in logic, mathematics, and the natural sciences is that any statement of the form "All A's are B's" is to be regarded as false *only* if there is at least one A who is not a B. And so the only way the statement, "All Frenchmen of the club wear berets," can be false is if there is at least one Frenchman in the club who doesn't wear a beret. If it so happens that there are no Frenchmen in the club, then there certainly isn't any Frenchman in the club who doesn't wear a beret, and therefore it is then taken as *true* that all the Frenchmen of the club wear berets. This is the convention we shall adopt.

Applying this to our universe U_3, if x is an individual with no ancestors, then anything one can say about *all* of his ancestors is true (because he has none!). In particular, given any proposition p, we take it as *true* that all of x's ancestors believe p, and so if x has no ancestors, then x believes Np. (The *only* way an individual x can fail to believe Np is if he has at least one ancestor who doesn't believe p, which is not possible for an individual who has no ancestors at all.) Let us record this as Fact 1.

Fact 1. If x has no ancestors, then for every proposition p, x believes Np.

We are aiming to show that the set of established propositions of U_3 is of type G. It certainly is of type 4 (by Problem 2, since all the conditions of U_2 hold also for U_3). It then remains for us to show that for any proposition p, all the inhabitants of U_3 believe the proposition $N(Np{\supset}p){\supset}Np$. The proof of this is very pretty; the key idea is contained in the following lemma.

Lemma 1. If x disbelieves Np, then x must have an ancestor y who both disbelieves p and believes Np.

First, prove the above lemma. Then show that the set of established propositions of U_3 is of type G.

How all this ties in with Kripke semantics will be explained in the next chapter.

SOLUTIONS

1 · (1) We know that conditions (1a) and (1b) are true, since each reasoner is of type 1.

(2) So next we must demonstrate that each reasoner x believes $(Np\&N(p{\supset}q)){\supset}Nq$, or, what is the same thing, if he believes

Np&N(p⊃q), then he must also believe Nq. So, suppose x believes Np&N(p⊃q). Then he believes both Np and N(p⊃q). Since he believes Np, then his parents both believe p. Since he believes N(p⊃q), then his parents both believe p⊃q. Therefore his parents both believe p and p⊃q, and being of type 1, they both believe q. Since his parents both believe q, then x believes Nq.

Thus we have proved that *every* reasoner x of U_2 believes (Np&N(p⊃q))⊃Nq, and so this proposition is established.

(3) Last, we must show that if p is established, so is Np. (This does not mean that every reasoner who believes p will also believe Np, but only that if *all* reasoners believe p, then they all believe Np.) This is really quite obvious. Suppose all reasoners believe p. Take any reasoner x. Then his parents believe p (because all the reasoners do), hence x believes Np.

2 · The importance of the change from "parents" to "ancestors" is this: If y is a parent of x, and z is a parent of y, we can hardly conclude that z is a parent of x. But if y is an *ancestor* of x, and z is an ancestor of y, then z is an ancestor of x. (In mathematical terminology, the relation of being an ancestor is *transitive*.)

The proof that the set of established propositions of the universe U_2 is of type 3 is the same as for the last universe U_1 (just changing the word "parent" to "ancestor"). But now we can prove the additional fact that every reasoner of this universe believes Np⊃NNp, and hence that the set of established propositions is of type 4.

Suppose x believes Np. We are to show that he must also believe NNp. Well, let x′ be any ancestor of x and let x″ be any ancestor of x′. Then x″ is also an ancestor of x. Since x believes Np and x″ is an ancestor of x, then x″ must believe p. Thus *every* ancestor x″ of x′ believes p, hence x′ believes Np. This shows that *every* ancestor x′ of x believes Np, and so x must believe NNp.

3 · First, to prove the lemma, suppose x disbelieves Np. Then he must have at least one ancestor x′ who disbelieves p (because if all

his ancestors believed p, he would believe Np, which he doesn't). Now, if x' believes Np, we are done—we take y to be x'. But if x' doesn't believe Np, then he disbelieves Np, hence x' must have at least one ancestor x" who disbelieves p. If x" believes Np, we are done (we take y to be x"). But if not, then we take some ancestor x'" of x" who disbelieves p, and we keep on going in this manner until we finally must reach some ancestor y of x who disbelieves p and who either has no ancestors at all (in which case y believes Np by Fact 1), or who does have ancestors all of whom believe p—and hence y must believe Np. (The reason we must finally reach such an ancestor y is because the universe U_3 had a definite beginning in time—a fact not given for the universe U_2!)

Now we can prove that all reasoners of this universe must believe N(Np⊃p)⊃Np (and hence that the set of established propositions is of type G). To prove this, it suffices to show that every reasoner who believes N(Np⊃p) will believe Np; or, what is the same thing, any reasoner who disbelieves Np will also disbelieve N(Np⊃p).

And so suppose that x disbelieves Np. Then by the lemma, x has an ancestor y who disbelieves p and believes Np. Then he believes Np true and p false, so he must disbelieve Np⊃p. Therefore x has an ancestor y who disbelieves Np⊃p, hence not all of x's ancestors believe Np⊃p, so x must *disbelieve* N(Np⊃p).

This proves that if x disbelieves Np, he disbelieves N(Np⊃p), hence if x believes N(Np⊃p), he must believe Np, and therefore x must believe N(Np⊃p)⊃Np.

· 23 ·

Possible Worlds

THE SUBJECT of modal logic is an ancient one—it goes back at least to Aristotle. The principal notions are that of a proposition being *necessarily* true and a proposition being *possibly* true. Either notion can be defined in terms of the other. If we start with the notion of necessary truth, we would define a proposition to be *possibly* true if it is not necessarily false. Alternatively, we could start out with the notion of a proposition being *possibly* true, and then define a proposition to be *necessarily* true if it is not possible that it is false.

Modal logic exercised considerable interest among the medieval philosophers and theologians and was later fundamental in the philosophy of Leibniz. It was Leibniz's thought that inspired the contemporary philosopher Saul Kripke to invent the field known today as *possible world semantics*, also called *Kripke semantics* (which is what we worked on in the last chapter, using a different terminology).

Leibniz had the idea that we inhabit a place called the actual world, which is only one of a number of *possible* worlds. According to Leibniz's theology, God first looked over *all* possible worlds and then actualized the one he thought best—this world. Hence his dictum: "This is the best of all possible worlds." (In *Candide*, Voltaire continually poked fun at this idea. After describing just

196

about every possible catastrophe, Voltaire would always add "—in this best of all possible worlds.")

To continue with Leibniz, a proposition p was to be thought of as true *for* or *in* a given world x (whether actual or possible) if it correctly described that world, and false for that world if it didn't. If p was called *true* without qualification, it was to be understood as meaning true for this (the actual) world. He called p *necessarily* true if it was true for *all* possible worlds, and *possibly* true if it was true for at least one possible world. Such—in brief—was the "possible world" philosophy of Leibniz. (If some other possible world had been actualized, I wonder if Leibniz would have had the same philosophy?)

Prior to 1910, the treatment of modal logic lacked the precision of other branches of logic. Even Aristotle, who was eminently clear in his theory of the syllogism, did not give an equally clear account of modal logic. It was the American philosopher C. I. Lewis who, in a series of papers published between 1910 and 1920, described a sequence of axiom systems of different strengths and investigated what propositions are provable in each. In each of these systems, all tautologies are among the axioms (or at least provable from them), and for any propositions p and q, Lewis took it as a rule that if p and p⊃q are both provable in the system, so is q. Thus all of Lewis's systems are at least of type 1. Next, Lewis reasoned that if p and p⊃q are both *necessarily* true, so is q. Hence all propositions of the form (Np&N(p⊃q))⊃Nq (or alternatively, all propositions of the form N(p⊃q)⊃(Np⊃Nq)) were taken as axioms. Next, it seemed reasonable that anything that could be *proved* purely on the basis of necessarily true axioms must be necessarily true, and today most modal systems (the so-called normal ones) take it as a rule of inference that if a proposition X has been *proved,* then we are justified in concluding NX. (This does *not* mean that X⊃NX is necessarily true, but that if X has been *proved*—purely on the basis of axioms that themselves are necessarily true—then we are justified in claiming NX.)

197

The system that we have described so far is of type 3 and has a standard name these days. It is called the *modal system K,* and is the basis of a wide class of modal systems.

Now, what about NX⊃NNX? (If X is necessarily true, is it *necessary* that it is necessarily true?) Well, propositions of this form are taken as axioms in some modal systems and not in others. The modal system whose axioms are those of K together with all propositions of the form NX⊃NNX is a very important one and is these days called the *modal system K_4.* It is obviously of type 4.

The modal system G—which consists of K_4 with the addition of all propositions of the form N(Np⊃p)⊃Np as axioms—came only decades later (in the mid-seventies) and did not arise out of any philosophical considerations of the notion of logical *necessity,* but out of Gödel's Second Theorem and Löb's Theorem. More about that in the next chapter.

Kripke Models. In the late 1950s, Saul Kripke published his famous paper, *A Completeness Theorem in Modal Logic,* which ushered in a new era for modal logic. For the first time a precise model theory was given for modal systems, which was not only of *mathematical* interest but which has led to a whole branch of philosophy known today as possible world semantics.

Kripke first raised a basic question about Leibniz's system that Leibniz evidently did not consider. According to Leibniz, we inhabit the actual world. Are the so-called possible worlds all the worlds that there are, or are they only those that are possible *relative to this world?* In other words, from the viewpoint of another world, is the class of possible worlds different from the class of worlds that are possible relative to this one? Or, to put the matter still another way, suppose we give a description of some world x, and we consider the proposition "x is a possible world." Is the truth or falsity of *that* proposition something absolute, or could it be that that very proposition is true in some world y but false in some other world z? Of particular importance here is the transitivity question: If world y is possible relative to world x and if world z is possible relative to world y, does it follow that world z must be possible relative to world x? The answer to that question determines which system of modal logic is appropriate.

Following Kripke, we shall say that world y is *accessible* to world x if y is possible relative to x. And now let us consider a "super-universe" of possible worlds. Given any worlds x and y, either y is accessible to x or it isn't. Once it is determined which worlds are accessible to which, we have what is technically called a *frame.* Given any proposition p and any world x, p is either true in x or false in x. And once it is determined which propositions are true in which worlds of the frame, we have what is called a *Kripke model.* It is to be understood that ⊥ is false in each of the worlds and that p⊃q is true in world x if and only if it is not the case that p is true in x and q is false in x. Thus, for each world x, the set of all propositions true in x is of type 1. To complete the description, a proposition Np is declared true *in world x* if and only if p is true *in all worlds accessible to x.* We will say that p is *established* in a model, or *holds* in a model, if p is true in *all* the worlds of the model.

The setup we now have is really the same as that of the last chapter, except for terminology. Instead of the elements of the universe being called *reasoners,* they are now called *worlds.* In place of the relation "y is a parent of x," or "y is an ancestor of x," we now say "y is accessible to x." Finally, in place of "p is believed by reasoner x," we now have "p is true in world x." With these transformations, all the results of the last chapter carry over. The set of all propositions that hold in a Kripke model must be of type 3 (Problem 1, Chapter 22), and hence the modal system K is applicable to all Kripke models.

Suppose we now add the transitivity condition—for any worlds x, y, and z, if y is accessible to x and z is accessible to y, then z is accessible to x. We then have what is called a *transitive* Kripke model. Well, for any transitive Kripke model, the set of all propositions that are true for all worlds of the model must be of type 4 (Problem 2, Chapter 22), and so the appropriate modal system is then applicable.

Thus the modal system K is applicable for all Kripke models, and the modal system K_4 is applicable to all *transitive* Kripke models. These two results are known as the *semantical soundness* theorems for K and K_4. Kripke also proved their converses: (1) If p holds in all Kripke models, then p is actually provable in the modal system K. (2) If p holds in all *transitive* Kripke models, then p is provable in K_4. (We will later explain exactly what is meant by *provability* in a modal system.) These two results are known as the *completeness* theorems for K and K_4.

Let us say that a Kripke model is *terminal* if the following condition holds: Given any world x of the model, if we pass to a world x' accessible to x and then to a world x" accessible to x', and keep going, we must finally reach a world y to which no worlds are accessible (so-called *terminal* worlds, which behave like the *parentless* reasoners of the last chapter). We shall say that a model is of type G if it is transitive and terminal. By the same reasoning as that of the last chapter, we see that the class of propositions that hold in a model of type G must be of type G, and hence the appropriate modal system is the modal system G. We thus get the so-called soundness theorem for the modal system G: Every proposition provable in G holds in all transitive terminal models. The converse of this —the completeness theorem for G—has also been established by the logician Krister Segerberg. It says that all propositions that hold in all models of type G are provable in the modal system G.[†]

[†]The proofs of the completeness theorems for the modal logics K, K_4, and G can be found in George Boolos, *The Unprovability of Consistency* (Cambridge University Press, 1979), and a greatly simplified proof for G can be found in George Boolos and Richard C. Jeffrey, *Computability and Logic* (Cambridge University Press, 1980; second edition).

As a philosophical analysis of the notion of *necessity,* the modal system G seems most inappropriate. Its real importance lies in the *provability* interpretation, which we will discuss in the next chapter.

Lewis's System S_4. Lewis had several other systems of modal logic, one of which we will briefly mention. Lewis reasoned that any proposition that is necessarily true must also be true. (In Leibniz's terms, if a proposition is true in *all* possible worlds, it should certainly be true in this one!) And so Lewis added as axioms to K_4 all propositions of the form NX⊃X. This is known as the *modal system S_4.*

The appropriate model theory for S_4 is a transitive Kripke model (but *not* a terminal one!) with the added condition that every world is accessible to itself. It is then easy to see that NX⊃X holds in such a model.

The modal systems S_4 and G form a genuine parting of the ways. It is impossible to combine the two systems into a single system without getting an inconsistent system (can the reader see why?). And so we must make a choice, depending on our purpose. As a philosophical analysis of the notion of necessary truth, the system S_4 seems the appropriate one. For proof theory, the system G is the important one. But more of this in the next chapter.

Exercise 1. Why is it impossible to combine the systems G and S_4 without getting an inconsistency?

Exercise 2. In a model of type G, *no* world can be accessible to itself. Why is this?

Exercise 3. Prove that in a model of type G, there is at least one proposition p and at least one world x such that the proposition Np⊃p is *false* in world x.

Exercise 4. Is the following true or false? In a Kripke model of type G, there is at least one world in which the crazy proposition N⊥ (⊥ is necessarily true) is actually true.

· 24 ·

From Necessity
to Provability

THE NEXT important development in modal logic after Kripke's modal semantics occurred in the early 1970s, when the *provability interpretation* was given to the word "necessary." It is surprising that this did not catch on generally sooner, since Gödel suggested it in a very short paper published in 1933. Gödel used the symbol "B" in place of Lewis's "N," and suggested that Bp be interpreted as p is *provable* (in the system of Arithmetic, or in any of the closely related systems investigated by Gödel). Now, these systems are all of type 4, and so the axioms of K_4 are all correct under that interpretation. However, the mathematical systems under investigation turned out to be even of type G (as discovered by Löb), hence it was only natural to invent a modal axiom system to take care of them. Thus the modal system G was born. It has been studied by several logicians, including Claudio Bernardi, George Boolos, D. H. J. de Jongh, Roberto Magari, Franco Montagna, Giovanni Sambin, Krister Segerberg, C. Smorynski, and Robert Solovay. Research concerning this system is still going on.

At this point, it will help to discuss modal axiom systems more rigorously. The symbolism of modal logic is that of propositional logic, with one new symbol added—we shall take this symbol to be "B." (We recall that Lewis used the symbol "N," and the more standard symbol these days is "□." But I prefer to use Gödel's symbol "B.")

By a *modal formula*—more briefly, a formula—we mean any expression formed according to the following rules:

(1) $\perp$ is a formula and each of the propositional variables p, q, r, . . . is a formula.

(2) If X and Y are formulas, so is (X⊃Y).

(3) If X is a formula, so is BX.

What we called in Chapter 6 (page 43) a *formula* could now be called a *propositional formula*. A propositional formula is a special case of a modal formula; it is one in which the symbol B does not occur. But we will be concerned from now on with modal formulas, and these will be called simply *formulas*.

The logical connectives $\sim$, &, v, ⊃, $\equiv$ are defined from ⊃ and $\perp$ in the manner explained in Chapter 8.

Each modal system has its own axioms. In each of the modal systems that we will consider, one starts from the axioms and successively proves new formulas by use of the following two rules:

Rule 1 (known as *modus ponens*). Having proved X and (X⊃Y), we can infer Y.

Rule 2 (known as *necessitation*). Having proved X, we can infer BX.

By a *formal proof* in the system is meant a nnite sequence of formulas (usually displayed vertically and read downward), called the *lines* of the proof, such that each line is an axiom of the system, or it comes from two earlier lines of the proof by Rule 1, or it comes from one earlier line of the proof by Rule 2. A formula X is called *provable* in the system if there is a formal proof whose last line is X.

The three systems that particularly interest us are the systems K, K_4, and G, whose axioms we review below.

Axioms of K: (1) All tautologies.

(2) All formulas of the form B(X⊃Y)⊃(BX⊃BY).

204

Axioms of K_4: Those of K together with
(3) All formulas of the form BX⊃BBX.

Axioms of G: Those of K_4 together with
(4) All formulas of the form B(BX⊃X)⊃BX.

Remarks. Let us refer to the axioms of group (4) as the *special* axioms of G. We recall the Kripke–de Jongh–Sambin theorem of Chapter 18, which is that if a system of type 3 can prove all sentences of the form B(BX⊃X)⊃BX, then it can also prove all sentences of the form BX⊃BBX. Thus we could have alternatively taken as our axioms for G those of groups (1), (2), and (4); the formulas of group (3) would then have been derivable. In other words, if we add the axioms of (4) to those of K, rather than K_4, we would still get the full modal system G. The system G is often presented in this alternative manner.

Discussion. Knowledge of these modal systems provides information about the more usual systems of mathematics. The modal system K holds good for any mathematical system S of type 3 (if we interpret BX as "X is provable in S"). Similarly, the axiom system K_4 holds good for any system S of type 4, and the axiom system G holds for any system S of type G. Thus these modal axiom systems give useful information about provability in the more common types of mathematical systems (which are nonmodal). Computer scientists today are also interested in modal axiom systems for the following reason. Imagine a computer programmed to print out various sentences, some of which are assertions about what the computer can and cannot print. The interpretation of BX then becomes: "The computer can print X." Such computers are, so to speak, "self-referential," and are accordingly of interest to those working in artificial intelligence. We will consider such systems in a later chapter.

In much of this book we have been treating "belief" as a modality.

205

We started out using "B" for "believed" (by a reasoner of appropriate type). Modal logic enables one to give a unified treatment of reasoners who *believe* propositions, computers that can *print* propositions (or rather, sentences that express them), and mathematical systems that can *prove* propositions.

Sentential Modal Systems. By a modal *sentence* we shall mean a modal formula in which none of the propositional variables p, q, r, . . . occur—expressions such as B⊥⊃⊥, B(⊥⊃B⊥). Thus modal sentences are all built from the five symbols B, ⊥, ⊃, (,). By a *sentential* modal system, we shall mean a modal system whose axioms are all sentences (from which it easily follows that only sentences are provable). For any modal system M, we shall let M̄ be that system whose axioms consist of all *sentences* that are axioms of M, and whose rules of inference are the same as those of M (usually they will be modus ponens and necessitation). We will be particularly interested in the sentential modal systems K̄, K̄₄, and Ḡ. It is not difficult to show that if M is either of these three systems, then any *sentence* provable in M is also provable in M̄. (The reader might try this now as an exercise—the solution will be given later—Chapter 27—when we need to use this fact.)

We will return to the study of modal systems after treating the fascinating topic of *self-reference*—to which we turn in the following chapter.

· Part X ·

THE HEART OF
THE MATTER

· 25 ·

A Gödelized Universe

L ET U S now turn to what can be called the heart of the matter
—namely, self-reference. We have not yet given the reader any idea
of *how* Gödel managed to construct a self-referential sentence—a
sentence that asserted its own nonprovability in the system under
consideration. He did this by inventing an extremely ingenious device known as *diagonalization.* In this chapter and the next, we will
consider Gödel's diagonal argument in several forms.

A GÖDELIZED UNIVERSE

Let us contemplate a universe with *infinitely* many reasoners in it.
There are also infinitely many propositions about this universe. More
specifically:

(1) ⊥ is one of these propositions (and, of course, is false).
(2) For any one of these propositions p and any reasoner R, the
proposition that R *believes* p is one of these propositions.
(3) For any propositions p and q about the universe, p⊃q is again
a proposition about the universe and is true if and only if either
p is false or q is true.

We henceforth use the word "proposition" to mean a proposition about the universe. We define the logical connectives $\sim$, &, v, $\equiv$ from $\supset$ and $\perp$ in the manner of Chapter 8.

For any reasoner R and any proposition p, we let Rp be the proposition that R believes p. For any reasoners R and S and any proposition p, RSp is the proposition that R believes that S believes p. If we throw in another reasoner K, KRSp is the proposition that K believes that R believes that S believes p—and similarly if we add more reasoners.

The reasoners had great fun reasoning about these propositions, but things were rather chaotic until a certain logician from another universe visited their universe and put things in order. The first thing he noticed was that the reasoners had no names, and so he assigned to each reasoner a number (a positive whole number, that is) known as the *Gödel number* of the reasoner. No two reasoners had the same number and every number was the number of some reasoner. Now the reasoners had names: R_1 is the reasoner whose number is 1, R_2 is the reasoner whose number is 2, and for each n, R_n is the reasoner whose number is n.

Next, the logician arranged all the propositions about the universe in a certain infinite sequence $p_1, p_2, \ldots, p_n, \ldots$ For each n, the number n was known as the Gödel number of the proposition p_n. After having done these things, the logician left and went back to his home universe.

Shortly after the logician's departure, the more clever of the inhabitants realized the following curious fact.

Fact 1. For each reasoner R, there was a reasoner R* such that for any proposition p_i, the reasoner R* believed p_i if and only if the reasoner R believed that R_i believed p_i. (Thus for any reasoner R, there was a reasoner R* such that for every number i, the proposition $R^*p_i \equiv RR_ip_i$ is true.)

This fact has some interesting consequences, as the following problems will reveal.

210

1 · The Diagonal Principle

Prove that for any reasoner R, there is at least one proposition p such that $p \equiv Rp$ is true (in other words, p is true if and only if R believes p).

2 · Inept Reasoners

A reasoner of this universe is called *totally inept* if he believes all false propositions and doesn't believe any true ones.

Prove that no reasoner of this universe is totally inept.

Another important fact about this universe was realized soon after the logician's departure.

Fact 2. For any reasoner R and any proposition q, there is some reasoner S such that for any proposition p, S believes p if and only if R believes $p \supset q$. (Thus $Sp \equiv R(p \supset q)$ is true.)

3 · A Tarskian Principle

A reasoner of this universe is called *perfect* if he believes all true propositions and doesn't believe any false ones. (He is the diametric opposite of a totally inept reasoner.)

For years the reasoners of this universe were in search of a perfect reasoner (of their own universe), but couldn't find one. Why were they unable to find one?

The next important things to be discovered about this universe are that certain propositions are called *established* and that there is a reasoner E who believes those and only those propositions that are established.

211

4

Assuming that all the established propositions are true, prove that the set of established propositions is incomplete—i.e., there must be at least one proposition p such that neither p nor $\sim$p is established. (This means also that the reasoner E can never believe p and can never believe $\sim$p. He must remain forever undecided as to whether or not p is true.)

Next, the following two facts were revealed.

Fact I. For any reasoner R, there is a reasoner R* such that for every number i, the proposition $R^*p_i \equiv RR_ip_i$ is *established*. (This differs from Fact 1 in that we now say *established* instead of *true.*)

Fact II. For any reasoner R and any proposition q, there is a reasoner S such that for every number i, $Sp_i \equiv R(p_i \supset q)$ is established. (This differs from Fact 2 in that we say established instead of true.)

5

Prove that for any reasoner R, there is a proposition p such that the proposition $p \equiv Rp$ is established.

6

Suppose that the set of established propositions is of type 1.

(a) Show that for every reasoner R and every proposition q, there is a proposition p such that the proposition $p \equiv R(p \supset q)$ is established.

(b) Show that for every reasoner R and every proposition q, there is a proposition p such that the proposition $p \equiv (Rp \supset q)$ is established.

Finally, the following fact was realized.

212

Fact III. The reasoner E was of type 4 (and hence the set of established propositions was of type 4).

7

Prove that the set of established propositions is of type G.

We now see that the set of established propositions of this universe is of type G, and hence, if this set is consistent, the *fact* that it is consistent, though true, is not one of the established propositions of the universe. Equivalently, if the reasoner E is consistent, he can never know that he is consistent.

RELATION TO MATHEMATICAL SYSTEMS

The reader might wonder what all this has to do with the theory of mathematical systems. Well, suppose we have a mathematical system S with all its propositions arranged in some infinite sequence p_1, $p_2, \ldots, p_n, \ldots$ Now, instead of *reasoners*, we will consider certain *properties* of propositions; these properties are also arranged in some infinite sequence $R_1, R_2, \ldots, R_n, \ldots$ For any *property* R_i and any proposition p_j, we now let $R_i p_j$ be the proposition that the property R_i *holds* for the proposition p_j. Suppose also that the property—call it E—of being a *provable* proposition of the system is one of the properties of the above list and suppose that Facts I, II, and III hold replacing the words "reasoner" by "property" and "established" by "provable" (provable in S, that is). Then by a mere change of terminology, the preceding arguments of this chapter show that the system S must be of type G.

Actually, the systems investigated by Gödel did not start out with properties of propositions, but with properties of *numbers*. However, by the device of Gödel-numbering the propositions, any property of numbers then corresponds to a certain property of propositions—

213

namely, for any property A of *numbers,* we let A' be the property which holds for just those *propositions* p_i such that A holds for the number i. A concrete illustration of this will be given in the next chapter.

Self-reference can also be achieved without Gödel numbering, as is indicated by the exercise below.

Exercise 1. There is another universe of reasoners in which it makes no difference whether the number of reasoners is finite or infinite. Some of the reasoners are immortal, but no reasoner knows which of the reasoners are immortal and which ones are mortal. In fact, no reasoner knows whether he himself is mortal or not. For every reasoner R, we let $\overline{R}$ be the proposition that R is immortal. For any reasoners R and S, we let $R\overline{S}$ be the proposition that R *believes* that S is immortal; for any three reasoners R, S, and K, we let $RS\overline{K}$ be the proposition that R believes that S believes that K is immortal —and so forth (if there are more reasoners involved).

In place of Fact 1 of the last universe, we have the following fact for this universe: For any reasoner R, there is a reasoner R^* such that for every reasoner S, the reasoner R^* believes that S is immortal if and only if R believes that S believes himself to be immortal (thus $R^*\overline{S}$ is true if and only if $RS\overline{S}$ is true).

Given a reasoner R, find a proposition p such that p is true if and only if R believes p.

Note: This method of achieving self-reference without Gödel numbering is borrowed from the field known as combinatory logic. A host of related self-referential problems in this field can be found in my book *To Mock a Mockingbird.*

SOLUTIONS

1 · Take any reasoner R. By Fact 1, there is a reasoner R^* such that for every number i, the reasoner R^* believes p_i if and only if R

believes the proposition $R_i p_i$. Now, R^* has some Gödel number h, and so R^* is the reasoner R_h. Thus for any number i, the following is true:

(1) R_h believes p_i if and only if R believes that R_i believes p_i.

Since this is true for *every* number i, it is also true when i is the number h. We thus have:

(2) R_h believes p_h if and only if R believes that R_h believes p_h.

We thus take p to be the proposition that R_h believes p_h, and we see that p is true if and only if R believes p.

2 · We have just seen that for any reasoner R, there is a proposition p such that p is true if and only if R believes p. This means that one of the following two cases must hold: (1) p is true and R believes p; (2) p is false and R doesn't believe p. If (1) holds, then R believes at least one true proposition—namely, p—and hence R is not totally inept. If (2) holds, then there is at least one false proposition—namely, p—such that R doesn't believe p, hence R doesn't believe *all* false propositions, and so again R is not totally inept.

3 · Using Fact 2, we take for q the proposition $\perp$. Then for any reasoner R, there is a reasoner R' (called "S," in Fact 2) who believes those and only those propositions p such that R believes $p \supset \perp$. (Such a reasoner R' might be said to *oppose* R.)

Now, suppose R were perfect. Then for any proposition p, R believes $p \supset \perp$ if and only if $p \supset \perp$ is true, which in turn is the case if and only if p is false. Therefore R believes $p \supset \perp$ if and only if p is false. Also, R' believes p if and only if R believes $p \supset \perp$. Putting these last two facts together, R' believes p if and only if p is false. This means that R' is totally inept.

We thus see that if the universe contains a perfect reasoner R, then it must also contain a totally inept reasoner R'. But we have proved in Problem 2 that the universe contains no totally inept reasoner. Therefore the universe contains no perfect reasoner.

4 · Let us call a reasoner *accurate* if he believes no false propositions. Consider now any accurate reasoner R. We saw in Problem 3 that none of the reasoners is perfect, hence R is also imperfect. This means that R either believes some false proposition, or he fails to believe some true proposition. Since R is accurate, he doesn't believe any false proposition, hence it must be that R fails to believe some true proposition. This proves that for any accurate reasoner R, there is at least one true proposition p that R fails to believe. Since p is true, $\sim$p is false, hence R, being accurate, doesn't believe $\sim$p either. Therefore, for every accurate reasoner R, his belief system is incomplete. There is at least one proposition p such that R neither believes p nor believes $\sim$p—he must remain forever undecided as to whether p is true or false.

Assuming that all the established propositions are true, the reasoner E is accurate (because he believes all the established propositions and no others). Therefore there is a proposition p such that E neither believes p nor believes $\sim$p, hence neither p nor $\sim$p is established.

5 · The proof is essentially that of Problem 1 using Fact I in place of Fact 1.

Given a reasoner R, there is a reasoner R_h (called R^*) such that for any number i, the proposition $R_h p_i \equiv RR_i p_i$ is established. Hence $R_h p_h \equiv RR_h p_h$ is established. Thus $p \equiv Rp$ is established, where p is the proposition $R_h p_h$.

6 · Suppose the set of established propositions is of type 1.

(a) Take any reasoner R and any proposition q. By Fact 2, there is a reasoner S such that for every p, the proposition $Sp \equiv R(p \supset q)$ is established. Also by Problem 5 (reading S for R), there is a proposition p such that $p \equiv Sp$ is established. It then follows that $p \equiv R(p \supset q)$ is established (since it is a logical consequence of the last two propositions).

(b) Again take any reasoner R and any proposition q. By (a), there is a proposition—call it p_1—such that $p_1 \equiv R(p_1 \supset q)$ is established. Hence $(p_1 \supset q) \equiv (R(p_1 \supset q) \supset q)$ is established (a trick we have used before), and so $p \equiv (Rp \supset q)$ is established, where p is the proposition $p_1 \supset q$.

7 · We are now given that E is of type 4, hence he is certainly of type 1. Then by (b) of the last problem, for any proposition q, there is a proposition p such that the proposition $p \equiv (Ep \supset q)$ is established —thus the set of established propositions is *reflexive*. And we know from Chapter 19 that any reflexive system of type 4 is of type G.

· 26 ·

Some Remarkable Logic Machines

FERGUSSON'S LATEST MACHINE

The logician Malcolm Fergusson of my book *The Lady or the Tiger?* was fond of constructing logic machines to illustrate important principles in logic and proof theory. One of his machines was described in that book. In Fergusson's later years, when he heard about Gödel's and Löb's theorems, he straightaway set out to construct a second machine, which he delighted in demonstrating to his friends. He proved to their satisfaction that the machine was a consistent and stable machine of type G, and he took particular delight in demonstrating that the machine, though consistent, could never prove its own consistency! The machine illustrates in a very simple and instructive manner the essential ideas behind Gödel's First and Second Incompleteness theorems as well as Löb's Theorem. I am accordingly happy to communicate the details to the reader.

The machine prints out various sentences built from seventeen symbols. The first seven symbols are the following:

P	$\perp$	$\supset$	$($	$)$	d	$,$
1	2	3	4	5	6	7

218

Underneath each of these seven symbols, I have written its *Gödel number*. The remaining ten symbols are the familiar digits 1, 2, 3, 4, 5, 6, 7, 8, 9, 0. These digits are assigned Gödel numbers as follows: The Gödel number of 1 is 89 (8 followed by one 9); the Gödel number of 2 is 899 (8 followed by two 9's); and so on, until 0, whose Gödel number is 89999999999 (8 followed by ten 9's). Thus each of the seventeen symbols has a Gödel number. Given a complex expression, one finds its Gödel number by replacing each symbol with its Gödel number—for example, the Gödel number of $(P\bot\supset\bot)$ is 412325. As another example, the Gödel number of P35 is 18999899999. For any expression E, by $\overline{E}$ we shall mean the Gödel number of E (written as a string of the digits 1, 2, ..., 0). Not every number is the Gödel number of an expression (for example, 88 is not the Gödel number of any expression). If n *is* the Gödel number of an expression, we shall sometimes refer to the expression as the nth expression. (For example, pd is the 16th expression; $\bot$ is the 2nd expression.)

The machine is *self-referential* in that the sentences printed by the machine express propositions about what the machine can and cannot print. An expression is called *printable* if the machine can print it. The symbol "P" means "printable," and for any expression E built from the seventeen symbols, if we want to write down a sentence that asserts that E is printable, we don't write down PE, but $P\overline{E}$ (i.e., P followed by the Gödel number of E). For example, a sentence that asserts that $(P\bot\supset\bot)$ is printable is $P\overline{(P\bot\supset\bot)}$—i.e., P412325.

For any expressions X and Y, Fergusson defined the *diagonalization of X with respect to Y* to be the expression $(X(\overline{X},\overline{Y})\supset Y)$. The symbol "d" abbreviates "diagonalization"—and for any expressions X and Y, the expression $Pd(\overline{X},\overline{Y})$ is a sentence expressing the proposition that the diagonalization of X with respect to Y is printable.

We shall now define what it means for an expression to be a *sentence* and what it means for a sentence to be *true*.

219

(1) ⊥ is a sentence and ⊥ is false.

(2) For any expression X, the expression P$\overline{X}$ is a sentence and is true if and only if the expression X is printable.

(3) For any expressions X and Y, the expression Pd($\overline{X},\overline{Y}$) is a sentence and is true if and only if the expression (X($\overline{X},\overline{Y}$)⊃Y)—which is the diagonalization of X with respect to Y—is printable.

(4) For any sentences X and Y, the expression (X⊃Y) is a sentence and is true if and only if either X is not true or Y is true.

It is to be understood that no expression is a sentence unless its being so is a consequence of the above rules. The logical connectives ~, &, v, ≡ are defined from ⊃ and ⊥ in the manner explained in Chapter 8.

Now we give the rules for what the machine can print. The machine is programmed to print out an infinite list of sentences sequentially. Certain sentences called *axioms* can be printed at any stage of the process. Among the axioms are all tautologies. (Thus for any tautology X, the machine can print X whenever it likes, regardless of what it has or has not printed at any previous stage.) Next, the machine is programmed so that for any sentences X and Y, if the machine has, at a certain stage, already printed X and X⊃Y, then it can print Y. Thus the machine is of type 1 (in the sense that the class of printable sentences is of type 1). Since it is true that if X and X⊃Y are both printable, so is Y, then the sentence (P$\overline{X}$&P($\overline{X⊃Y}$))⊃P$\overline{Y}$ is true; or, what is the same thing, the sentence P($\overline{X⊃Y}$)⊃(P$\overline{X}$⊃P$\overline{Y}$) is true (the two sentences are logically equivalent). Well, the machine "knows" the truth of all sentences of the form P($\overline{X⊃Y}$)⊃(P$\overline{X}$⊃P$\overline{Y}$) and takes them all as axioms. Thus the machine is of type 2. Next, if the machine ever prints a sentence X, it "knows" that it has printed X and will sooner or later print the true sentence P$\overline{X}$. (The sentence P$\overline{X}$ *is* true, since X has been printed.) And so the machine is normal, hence is of type 3. Since the machine is normal, then for any sentence X, the sentence P$\overline{X}$⊃PP$\overline{X}$ is true. So the machine is initially "aware" of the truth of

220

all such sentences and takes them all as axioms. Thus the machine is of type 4.

There is one more thing the machine can do, and this is quite crucial. For any expressions X and Y, the sentence $Pd(\overline{X},\overline{Y})$ is true if and only if $(X(\overline{X},\overline{Y})\supset Y)$ is printable, which in turn is the case if and only if the sentence $P\overline{(X(\overline{X},\overline{Y})\supset Y)}$ is true. Therefore the following sentence is true: $Pd(\overline{X},\overline{Y})\equiv P\overline{(X(\overline{X},\overline{Y})\supset Y)}$.

Well, the machine knows the truth of all such sentences and takes them all as axioms. These axioms are called the *diagonal axioms.*

Let us now systematically review all the axioms and operations of the machine:

Axioms: Group 1. All tautologies.

Group 2. All sentences of the form $P\overline{(X\supset Y)}\supset(P\overline{X}\supset P\overline{Y})$.

Group 3. All sentences of the form $P\overline{X}\supset PP\overline{X}$.

Group 4 (the diagonal axioms). All sentences of the form $Pd(\overline{X},\overline{Y})\equiv P\overline{(X(\overline{X},\overline{Y})\supset Y)}$, where X and Y are any expressions (not necessarily sentences).

Operation Rules. (1) Axioms can be printed at any stage.

(2) Given sentences X and $(X\supset Y)$ already printed, the machine can then print Y.

(3) Given a sentence X already printed, the machine can print $P\overline{X}$.

This concludes the rules governing printability by the machine. It is to be understood that the only way the machine can print a sentence X at a given stage is by following one of the above rules. That is, X is printable at a given stage *only* if one of the following three conditions holds: (1) X is an axiom; (2) there is a sentence Y such that Y and $(Y\supset X)$ have both been printed at an earlier stage; (3) there is a sentence Y such that X is the sentence $P\overline{Y}$ and Y has been printed at an earlier stage.

Remarks. For each sentence X, let BX be the sentence $P\overline{X}$. The symbol "B" is *not* part of the machine language; we are using it to *talk about* the machine. We are using "B" as standing for the operation that assigns to each sentence X the sentence $P\overline{X}$. When we say that the machine is of type 4, we mean that it is of type 4 with reference to this Operation B. In essence, without the diagonal axioms, the axiom system of this machine is the modal system K_4. We will shortly see that adding the diagonal axioms gives us all the power of the modal system G.

Provability. We have defined for each sentence what it means for the sentence to be *true*, and so each sentence expresses a definite proposition, which might be true or might be false. We say that the machine *proves* a proposition if it prints some sentence that expresses the proposition. For example, the sentence $\sim P2$ expresses the proposition that the machine is consistent (since 2 is the Gödel number of $\bot$), and so if the machine printed $\sim P2$, it would prove its own consistency. If the machine printed P2, then it would prove its own *inconsistency*.

We say that the machine is *accurate* if all propositions provable by the machine are true. We say that the machine is *consistent* if it cannot prove $\bot$, and that the machine is *stable* if for every sentence X, if $P\overline{X}$ is printable, so is X.

Reflexivity. Now we turn to the proof that the machine is Gödelian —in fact, reflexive.

1 · The Gödel Sentence G

Find a sentence G such that the sentence $G \equiv \sim P\overline{G}$—i.e., the sentence $G \equiv (P\overline{G} \supset \bot)$—is printable.

2 · Reflexivity

Show that for any sentence Y, there is a sentence X such that the sentence $X \equiv (P\overline{X} \supset Y)$ is printable.

Solutions. Problem 1 is a special case of Problem 2, so we first solve Problem 2. Let Y be any sentence. For any expression Z, the sentence $Pd(\overline{Z},Y) \equiv P(\overline{Z(\overline{Z},Y) \supset Y})$ is printable (because it is one of the diagonal axioms). We take for Z the expression Pd, and therefore $Pd(\overline{Pd},Y) \equiv P(\overline{Pd(\overline{Pd},Y) \supset Y})$ is printable. Since the machine is of type 1, it then follows that the following sentence is printable: $(Pd(\overline{Pd},Y) \supset Y) \equiv (P(\overline{Pd(\overline{Pd},Y) \supset Y}) \supset Y)$.

Thus the sentence $X \equiv (P\overline{X} \supset Y)$ is printable, where X is the sentence $(Pd(\overline{Pd},Y) \supset Y)$.

Problem 1 is a special case of 2, taking $\perp$ for Y. Thus the Gödel sentence G for this machine is $Pd(\overline{Pd},\perp) \supset \perp$—i.e., the sentence $(Pd(16,2) \supset \perp)$.

Let us take a closer look at Gödel's remarkable sentence G. First, what does the sentence $Pd(16,2)$ say? It says that the diagonalization of the 16th expression with respect to the 2nd expression is printable. Now, the 16th expression is Pd and the 2nd expression is $\perp$, and so $Pd(16,2)$ says that the diagonalization of Pd with respect to $\perp$ is printable, but this diagonalization is the sentence $(Pd(16,2) \supset \perp)$—i.e., the very sentence G! And so $Pd(16,2)$ says that G *is* printable, hence $(Pd(16,2) \supset \perp)$—which is the sentence G—says that G is *not* printable (or, what is the same thing, that the printability of G implies falsehood). Thus G says that G is not printable; G is *true* if and only if G is not printable. Thus G asserts its own nonprintability. Here, in a nutshell, is Gödel's ingenious method of achieving self-reference.

The sentence $G \equiv \sim P\overline{G}$—i.e., the sentence $G \equiv (P\overline{G} \supset \perp)$—is not only true, but actually printable (it is one of the diagonal axioms). Since the machine is normal and of type 1, it follows by Gödel's First Incompleteness Theorem (Theorem I, Chapter 20, page 174) that if the machine is consistent, then G is not printable, and if the machine is also stable, then $\sim$G is also not printable. And so, *if* the machine is both consistent and stable, the sentence G is undecidable in the system of sentences that the machine can print.

Now, the machine is in fact of type 4, and since it is Gödelian

—the sentence $G \equiv \sim P\overline{G}$ is printable—it then follows from Gödel's Second Incompleteness Theorem (Part 4 of Summary I*, Chapter 13, page 110) that if the machine is consistent, then it cannot prove its own consistency—i.e., it cannot print the sentence $\sim P2$. Also, if the machine is consistent, then the sentence $\sim P2$ is *true*, and is hence another example of a true sentence that the machine cannot print.

Furthermore, the machine is reflexive (Problem 2) and being of type 4, it must be Löbian (by Löb's Theorem), and so for any sentence X, if $P\overline{X} \supset X$ is printable, so is X. It then follows by Theorem M_1, Chapter 18 (page 157), that the machine is of type G.

THE CORRECTNESS OF THE MACHINE

We have shown that *if* Fergusson's machine is consistent, then it cannot prove its own consistency; but how do we know whether or not the machine is consistent? We will now prove that the machine is not only consistent, but wholly accurate—i.e., that every sentence printed by the machine is true.

We have already shown that all the *axioms* of the machine are true, but let us carefully review the arguments: The axioms of Group 1 are all tautologies, hence they are certainly true. As for the axioms of Group 2, to say that $P(\overline{X \supset Y}) \supset (P\overline{X} \supset P\overline{Y})$ is true is to say that if $P(\overline{X \supset Y})$ and $P\overline{X}$ are both true, so is PY—which is to say that if $(X \supset Y)$ and X are both printable, so is Y. Well, this is obviously the case by virtue of Operation 2. Thus the axioms of Group 2 are all true. As for the axioms of Group 3, to say that $P\overline{X} \supset PP\overline{X}$ is true is to say that if $P\overline{X}$ is true, so is $PP\overline{X}$, which in turn is to say that if X is printable, so is $P\overline{X}$—and this is indeed the case, by virtue of Operation 3. As for the diagonal axioms, $Pd(\overline{X},\overline{Y})$ is true if and only if $(X(\overline{X},\overline{Y}) \supset Y)$ is printable, which is the case if and only if $P(\overline{X(\overline{X},\overline{Y}) \supset Y})$ is true. Therefore $Pd(\overline{X},\overline{Y}) \equiv P(\overline{X(\overline{X},\overline{Y}) \supset Y})$ is true.

Now we know that all the axioms of the machine are true, but we need to show that all the *printable* sentences are true.

We recall that the machine prints sentences at various *stages*. We now wish to establish the following lemma, theorem, and corollary.

Lemma. If X is a sentence printed at a certain stage and all sentences printed prior to that stage are true, then X is also true.

Theorem 1. Every sentence printed by the machine is true.

Corollary. The machine is both consistent and stable.

3

How are the above lemma, theorem, and corollary proved?

Solutions. First we prove the lemma: Assume the hypothesis that all sentences previously printed are true; we are to show that X is true.

Case 1. X is an axiom. Then X is true (as we have already proved).

Case 2. There is a sentence Y such that Y and $(Y \supset X)$ have already been printed. Then by the assumed hypothesis, Y and $(Y \supset X)$ are both true, hence X must be true.

Case 3. X is of the form $P\overline{Y}$, where Y is a sentence that has been previously printed. Since Y has been printed, then $P\overline{Y}$ is true—i.e., X is true.

This concludes the proof of the lemma.

Proof of Theorem 1. The machine is programmed to print all printable sentences in some definite infinite sequence X_1, X_2, ..., X_n, ... By X_n is meant the sentence printed at stage n. Now, the first sentence printed by the machine (the sentence X_1) must be an axiom (since the machine hasn't printed any other sentences yet), hence X_1 must be true. If the above infinite list should contain

225

any false sentence, then there must be a *smallest* number n such that X_n is false—that is, there must be a *first* false sentence that the machine prints. We know that n is not equal to 1 (since X_1 is true), hence n is greater than 1. This means that the machine prints a false sentence at stage n but has printed only true sentences at all earlier stages. But this is contrary to the lemma. Therefore the machine can never print any false sentences.

Proof of Corollary 1. Since the machine is accurate (by Theorem 1), then ⊥ can never be printed, because ⊥ is false. Therefore the machine is consistent.

Next, suppose that $P\overline{X}$ is printable. Then $P\overline{X}$ is true (by Theorem 1), which means that X is printable. Therefore the machine is stable.

We now see that Fergusson's machine *is* consistent, but can never prove its own consistency. Thus you and I (as well as Fergusson) know that the machine is consistent, but the poor machine doesn't have that knowledge!

CRAIG'S VARIANT

When Inspector Craig (a good friend of Fergusson's) heard of Fergusson's machine, he thought of an interesting variant, which does not involve Gödel numbering. Craig's machine used just the following six symbols:

P ⊥ ⊃ ( ) R

His definitions of *sentence* and *true sentence* are given by the following rules:

(1) ⊥ is a sentence and ⊥ is not true.
(2) For any sentences X and Y, (X⊃Y) is a sentence and is true if and only if X is not true or Y is true.

(3) For any sentence X, PX is true if and only if the machine can print X.

(4) For any sentences X and Y, the expression (XRY) is a sentence and is declared true if and only if the machine can print ((XRX)⊃Y).

At first sight it might appear that (4) is circular, since the truth of (XRY) is defined in terms of an expression involving the letter R; but this circularity is only apparent. If Craig had defined (XRY) to be true if and only if ((XRX)⊃Y) is *true,* the definition would have been circular, since we couldn't know what it means for (XRY) to be true without first knowing what it means for (XRX) to be true. But Craig didn't do that. The fact is that for any sentences X and Y, either the expression ((XRX)⊃Y) is printable or it isn't. The symbol "R" stands for the relation that holds between X and Y when ((XRX)⊃Y) *is* printable. Thus Craig is not defining the relation R in terms of itself, but in terms of the *symbol* "R," and this constitutes no circularity.

The first three groups of axioms of Craig's machine are the same as those of the modal system K_4 (where *sentence* means sentence in the machine language of Craig's system). Thus they are like the first three groups of axioms of Fergusson's machine, leaving out the bars over X, Y, Z. The fourth group of Craig's axioms can be read as: (4)' (Craig's diagonal axioms). All sentences of the form (XRY)≡P((XRX)⊃Y).

Of course all of Craig's diagonal axioms are true sentences.

The operation rules of Craig's machine are the same as those of Fergusson's machine.

4

(a) Prove that Craig's machine is reflexive (and hence also of type G, since it is of type 4).

(b) Find a Gödel sentence for Craig's machine (i.e., a sentence G such that G≡~PG is printable by Craig's machine).

Solution. (a) Since for any sentences X and Y, the sentence (XRY)≡P((XRX)⊃Y) is printable (it is a diagonal axiom), then this is also the case if X happens to be the very sentence Y. Therefore (YRY)≡P((YRY)⊃Y) is printable, hence so is the sentence ((YRY⊃Y)≡(P(YRY)⊃Y)⊃Y), and thus Z≡(PZ⊃Y) is printable, where Z is the sentence ((YRY)⊃Y).

(b) Taking ⊥ for Y, we get the Gödel sentence ((⊥R⊥)⊃⊥).

Note: The axioms of Craig's machine are also all true, and by the same reasoning we used for Fergusson's machine, it can be seen that every sentence printable by Craig's machine is true. Therefore Craig's machine is also consistent and stable (and of type G).

5 · McCulloch's Observation

When Walter McCulloch (a friend of both Craig and Fergusson) was informed about Craig's machine, he made the following interesting observation: Given any sentences X and Y in which the symbol "R" does not occur, there is a sentence Z in which "R" does not occur such that the sentence (XRY≡Z) is printable. (This implies, incidentally, that for any sentence X at all, there is a sentence X' in which "R" does not occur such that X≡X' is printable by Craig's machine.)

Can you prove that McCulloch's observation is correct?

Solution. We proved in the solution of Problem 8 of Chapter 19 that any system of type G which can prove p≡B(p⊃q) can prove p≡Bq. (We showed this for reasoners, but the same argument goes through for systems.) Now, Craig's system is of type G, and for any sentence X, the sentence (XRX)≡P((XRX)⊃X) is printable, and so if we take (XRX) for p and X for q, it follows that (XRX)≡PX is printable. From this it follows that ((XRX)⊃Y)≡(PX⊃Y) is printable, and hence (by regularity) P((XRX)⊃Y)≡P(PX⊃Y) is printable. But also (XRY)≡P((XRX)⊃Y) is printable, hence (XRY)≡P(PX⊃Y) is printable. If now the symbol "R" doesn't occur in either X or Y, then it doesn't occur in P(PX⊃Y), and so we take Z to be P(PX⊃Y).

· 27 ·

Modal Systems
Self-Applied

INSPIRED BY the logic machines of Craig and Fergusson, I would like now to look at modal axiom systems from the viewpoint of *self-referential interpretations.*

We recall that by a *modal sentence*—more briefly, a *sentence*—we mean a modal formula without propositional variables. We now define a modal sentence to be *true* for a modal system M if it is true when we interpret B as *provable in M.* Thus:

(1) $\perp$ is false for M.

(2) For any modal sentences X and Y, the sentence X⊃Y is true for M if and only if either X is *not* true for M or Y *is* true for M.

(3) For any sentence X, the sentence BX is true for M if and only if X is *provable* in M.

Note: A sentence X can be *true* for a modal system M without being provable in M, and conversely. For example, the sentence ∼B⊥ is *true* for M if and only if M is consistent. To say that ∼B⊥ is *provable* in M is to say that M can prove its own consistency. We will soon see that the modal system G is consistent, and so the sentence ∼B⊥ is *true* for G. But the sentence ∼B⊥ is not *provable* in G. On the other hand, any inconsistent modal system of type 1 can prove all sentences, hence in particular the sentence ∼B⊥. In this case the sentence ∼B⊥ is *provable* in the system, but not *true* for the system.

Consider now a modal system M_1 and a modal system M_2 (possibly the same as M_1 or possibly different). We shall say that M_1 is *correct* for M_2 if every sentence provable in M_1 is *true* for M_2. And we shall say that a modal system M is *self-referentially correct* if M is correct for M—in other words, if every sentence provable in M is true for M.

Any self-referentially correct system must be consistent (because if ⊥ were provable in the system, the system couldn't be self-referentially correct, since ⊥ is false for the system) and must also be stable (because if BX is provable in the system and the system is self-referentially correct, then BX must be true for the system, which means that X is provable in the system). And so any self-referentially correct system is automatically both consistent and stable.

Self-referential correctness has one curious feature. It is possible that a system M might be self-referentially correct; yet if some of the axioms were deleted, the resulting system might no longer be self-referentially correct. For example, we might have one axiom that asserts that a second axiom is provable in the system; if this second axiom is removed, the first axiom might become false!

Some Self-Referentially Correct Systems. We recall the sentential modal systems $\overline{K}$, $\overline{K_4}$, and $\overline{G}$ described in Chapter 24 (they are like the systems K, K_4, and G, except that the axioms are all restricted to sentences). We now aim to show that these three systems are self-referentially correct (from which, incidentally, we will be able to show that the systems K, K_4, and G are self-referentially correct).

If $\overline{M}$ is any of these three axiom systems $\overline{K}$, $\overline{K_4}$, $\overline{G}$, to show that $\overline{M}$ is self-referentially correct, it suffices to show that all *axioms* of $\overline{M}$ are true for $\overline{M}$—the reason for this is a consequence of the following lemma, which has other applications as well.

Lemma A. Let $\overline{M_1}$ be any sentential modal system whose only inference rules are modus ponens (from X and X⊃Y we can infer Y) and the necessitation rule (from X we can infer BX). Let M_2 be any modal system such that all sentences provable in $\overline{M_1}$ are provable in M_2 and such that all *axioms* of $\overline{M_1}$ are *true* for M_2. Then all sentences provable in $\overline{M_1}$ are true for M_2—i.e., $\overline{M_1}$ is *correct* for M_2.

Corollary A_1. For any sentential modal system $\overline{M}$ whose only inference rules are modus ponens and necessitation, if all *axioms* of $\overline{M}$ are true for $\overline{M}$, then $\overline{M}$ is self-referentially correct.

1

Prove Lemma A. (Hint: The proof is essentially the same as the argument we gave in the last chapter to show that all provable sentences of Fergusson's machine were true—once we established that all the axioms of the machine were true.)

Solution. Consider any sequence $X_1, \ldots, X_n$ of sentences that constitutes a proof in the system $\overline{M_1}$. We will see that the first line X_1 must be true for M_2, then the second line X_2 must be true for M_2, then the third line X_3, and so forth down to the last line.

The first line X_1 must be an axiom, hence it is true for M_2 by hypothesis. Now consider the second line X_2. Either it is an axiom of $\overline{M_1}$ (in which case it is true for M_2), or it must be the sentence BX_1, in which case it is certainly true for M_2 since X_1 has already been proved in M_1, and is hence provable in M_2. Now we know that the first two lines are true for M_2. We now consider the third line X_3. If it is either an axiom of $\overline{M_1}$ or is of the form BY, where Y is an earlier line (X_1 or X_2), then X_3 is true for M_2 for the same reasons as before. If X_3 is neither, then it must be derived from X_1 and X_2 by modus ponens, and since we already know that X_1 and X_2 are true for M_2, it follows that X_3 is true for M_2. (Clearly, for any sentences X and Y, if X and X⊃Y are both true for M_2, then Y is true for M_2.) Now we know that the first three lines of the proof are all true for M_2, and knowing this, the truth of X_4 can be established by the same argument. Then, knowing the truth of the first four lines, we similarly get the truth of the fifth line—and so on, until we reach the last line.† This concludes the proof of Lemma A.

The self-referential correctness of the systems $\overline{K}$, $\overline{K_4}$, and $\overline{G}$ follows from Corollary A_1 and the following lemma.

Lemma B. For any modal system M:

(a) If M is of type 1, then all axioms of $\overline{K}$ are true for M.

(b) If M is normal and of type 1, then all axioms of $\overline{K_4}$ are true for M.

(c) If M is of type G, then all axioms of $\overline{G}$ are true for M.

2

Why is Lemma B correct?

Solution. (a) Suppose that the set of provable sentences of M is closed under modus ponens. All tautologies are obviously true for M (they are true for any modal system whatsoever). The other axioms of $\overline{K}$ are sentences of the form $(BX\&B(X{\supset}Y)){\supset}BY$—or alternatively $B(X{\supset}Y){\supset}(BX{\supset}BY)$; it really makes no difference. To say that $(BX\&B(X{\supset}Y)){\supset}BY$ is true for M is to say that if X is provable in M and $X{\supset}Y$ is provable in M, so is Y. Well, this is the case, since the provable sentences of M are closed under modus ponens.

(b) Suppose M is a normal system of type 1. Then all axioms of $\overline{K}$ are true for M—by (a). The other axioms of $\overline{K_4}$ are the sentences of the form $BX{\supset}BBX$. Well, to say that such a sentence is true for M is to say that if BX is true for M, so is BBX—in other words, if X is provable in M, so is BX. This is so, since M is normal.

(c) Suppose M is of type G. Then it is certainly of type 4, so by (b), all axioms of $\overline{K_4}$ are true for M. The remaining axioms of $\overline{G}$ are

---

†This type of argument is known as *mathematical induction*.

sentences of the form B(BX⊃X)⊃BX, and to say that it is *true* for M is to say that if BX⊃X is provable in M, so is X—in other words, that M is Löbian. Well, we proved in Chapter 19 that any system of type G is Löbian. This concludes the proof of Lemma B.

We continue to let M be any of the systems K, K_4, or G. Then $\overline{M}$ is respectively $\overline{K}$, $\overline{K_4}$, or $\overline{G}$.

Corollary B_1. All axioms of $\overline{M}$ are true for $\overline{M}$.

Corollary B_2. All axioms of $\overline{M}$ are true for M.

Proofs. Since $\overline{K}$, $\overline{K_4}$, and $\overline{G}$ are respectively of type 1, normal and of type 1, and of type G, Corollary B_1 is immediate from Lemma B. Also the systems K, K_4, and G are respectively of type 1, normal and of type 1, and of type G, and so we also have Corollary B_2.

From Corollary B_1 and Corollary A_1, we now have Theorem 1.

Theorem 1. The systems $\overline{K}$, $\overline{K_4}$, and $\overline{G}$ are all self-referentially correct.

Corollary. The systems $\overline{K}$, $\overline{K_4}$, and $\overline{G}$ are consistent and stable.

We now have a third example of a consistent and stable system of type G—namely, the modal system $\overline{G}$ (the other two systems being the machines of the last chapter). We will soon see that the modal system G is also self-referentially correct (hence also both consistent and stable).

I hope the reader now fully realizes the absurdity of doubting the consistency of a system on the mere grounds that it cannot prove its own consistency!

The Systems K, K_4, and G. To establish the self-referential correctness of the systems K, K_4, and G, we proceed as follows. First of all, Lemma A has another corollary.

Corollary A_2. Let M be any modal system whose only inference rules are modus ponens and necessitation. Then, if all axioms of $\overline{M}$ are true for M, all provable sentences of $\overline{M}$ are true for M.

It follows from Corollary A_2 and Corollary B_2 that if M is any of the modal systems K, K_4, or G, then all sentences provable in $\overline{M}$ are true for M. But we are not quite done. It remains to show that if M is either of these three systems, then any sentence provable in M is also provable in $\overline{M}$ (a fact that was asserted without proof at the end of Chapter 19). Once this is done, the proof of the self-referential correctness of K, K_4, and G will be complete.

3

Why is it true that if a sentence is provable in M (M being either K, K_4, or G), then it is provable in $\overline{M}$?

Solution. One can see by inspection of either of these three systems that if X is an axiom of M, then if we substitute any sentences for the propositional variables in X (substituting the same sentence for different occurrences of the same variable, of course), then the resulting sentence is also an axiom of M—and hence of $\overline{M}$. We now take any one particular sentence, say T, and for any formula X, let X′ be the result of substituting T for *all* the propositional variables in X. Of course if X is itself a sentence, we take X′ to be X. We now note the following facts: (1) If X is an axiom of M, then X′ is an axiom of $\overline{M}$. (2) For any formulas X and Y, the sentence $(X{\supset}Y)'$ is the sentence $X'{\supset}Y'$, and therefore for any formulas X, Y, and Z, if Z is derivable from X and Y by modus ponens, then Z′ is derivable from X′ and Y′ by modus ponens. (3) For any formula X, the sentence $(BX)'$ is the sentence BX' (i.e., B followed by X′), and so if Y is derivable from X by the necessitation rule, then Y′ is derivable from X′ by the necessitation rule. It therefore follows that given any sequence $X_1, \ldots, X_n$ of formulas, if this sequence constitutes a proof in the system M, then the sequence X_1', X_2', ..., X_n' constitutes a proof in $\overline{M}$. And so, if X is any formula provable in M, the sentence X' is then provable in $\overline{M}$. If, furthermore, X happens to be a sentence, then $X' = X$, and hence X itself is provable in $\overline{M}$. This proves that any sentence provable in M is provable in $\overline{M}$.

· *Part XI* ·

FINALE

· 28 ·

Modal Systems, Machines, and Reasoners

IN THE next chapter we will meet up with some very strange reasoners indeed. To appreciate them fully, let us first turn to the topic of minimal reasoners.

MINIMAL REASONERS OF VARIOUS TYPES

A modal sentence X is in itself neither true nor false; it only expresses a definite proposition once the symbol "B" is given an interpretation. We have defined a sentence to be *true* for a modal system M if it is true when "B" is interpreted as provability in M. We have all along understood a modal sentence as being true for a *reasoner* if it is true when "B" is interpreted as *believed by the reasoner.*

To say that a sentence is *true* for a reasoner means something entirely different than saying that it is *believed* by the reasoner. For example, to say that ~B⊥ is true for a reasoner is to say that the reasoner is consistent, whereas to say that ~B⊥ is believed by a reasoner is to say that the reasoner believes that he is consistent.

239

A machine can easily be programmed to print out all and only those sentences provable in the modal system $\overline{G}$ by giving the machine these instructions: (1) At any stage, you may print any axiom of $\overline{G}$. (2) If at any stage you have printed sentences X and (X⊃Y), you may then print Y. (3) If at any stage you have printed X, you may then print BX. (The machine can then be given further instructions that will guarantee that anything the machine *can* do, it sooner or later *will* do, and so every sentence provable in $\overline{G}$ will eventually be printed by the machine.) Let us call such a machine a $\overline{G}$ machine.

Now, let us imagine a reasoner with his eye constantly on the output of the machine. However, he does not interpret BX as "X is printable by the machine," or as "X is provable in $\overline{G}$," but as "I believe X." (He thinks that the machine is printing sentences about *him!*) He gives what we might call an *egocentric* interpretation of modal sentences.

Then, suppose that the reasoner has complete confidence that the machine knows what he believes, and so whenever the machine prints a sentence X, he immediately believes it (under the egocentric interpretation, of course). His belief system will then include *all* sentences provable in $\overline{G}$. This does *not* guarantee that he is of type G (he may not be normal, even though he believes he is, and he may not even be of type 1, even though he believes he is). Of course if he *correctly* believes all sentences provable in $\overline{G}$, then it is easy to see that he is of type G.

But now, suppose he believes all *and only those* sentences printable by the machine; his belief system then coincides exactly with the set of sentences provable in $\overline{G}$, and since $\overline{G}$ is of type G, then *he* must be of type G. Such a reasoner we will call a *minimal* reasoner of type G. Since the system $\overline{G}$ is self-referentially correct (as we showed in the last chapter), it follows that a minimal reasoner of type G must be wholly accurate in his beliefs. It further follows that any minimal reasoner of type G is both consistent and stable.

We now see that the notion of a consistent, stable reasoner of type G does *not* involve a logical contradiction. A reasoner of type G is not necessarily consistent (indeed, any *inconsistent* reasoner of even type 1 is also of type G, since he believes everything!), but a *minimal* reasoner of type G is both consistent and stable.

Now let us consider a reasoner who *is* of type G and who is keeping his eye on the output of the machine (and who interprets all the modal sentences egocentrically). Will he necessarily *believe* all these sentences (under the egocentric interpretation)? Well, it is easy to verify that he believes all *axioms* of $\overline{G}$. (In fact, in Chapter 11 we showed that any reasoner of type 4 knows that he is of type 4; hence he will believe all axioms of $\overline{K}_4$. A reasoner of type G also believes he is modest, which means that he will believe all sentences of the form B(BX⊃X)⊃BX, and so he believes all axioms of $\overline{G}$.) And since the machine prints nonaxioms only by using the modus ponens and necessitation rules, and the reasoner's beliefs are closed under modus ponens, and he is normal, then he will successively believe every sentence as it gets printed. (If anyone should interrupt the process and ask the reasoner his opinion of the machine, the reasoner will answer: "This machine is truly amazing. Everything it has printed about me so far is true!")

We now see that a reasoner of type G does indeed believe all sentences provable in $\overline{G}$.

Suppose, next, that a modal sentence X is believed by all reasoners of type G; does it follow that X is actually provable in $\overline{G}$? The answer is yes, since if X is believed by *all* reasoners of type G, then it must be believed by a *minimal* reasoner of type G, and hence must be provable in $\overline{G}$. And so we now see that a sentence is provable in $\overline{G}$ *if and only if* it is believed by all reasoners of type G. Put another way, given any minimal reasoner of type G, he believes those and only those sentences that are believed by *all* reasoners of type G.

Of course, when two different reasoners look at the same modal sentence, they interpret it differently—each one interprets "B" as referring to his *own* beliefs (just as the word "I" has different references when used by different people). And so when we speak of a modal sentence being *believed* by all reasoners of type G, we

mean believed by each one according to his own egocentric interpretation.

Of course everything we have said about the modal system $\overline{G}$ and
reasoners of type G also holds for the modal system $\overline{K_4}$ and reasoners
of type 4: A sentence is provable in $\overline{K_4}$ if and only if it is believed
by all reasoners of type 4. Likewise, a sentence is provable in $\overline{K}$ if
and only if it is believed by all reasoners of type 3.

MORE ON MODAL SYSTEMS AND REASONERS

The results of the problems in the remainder of this chapter are not
necessary for the understanding of the next two chapters, but are
independently interesting.

1

Suppose a reasoner's beliefs are closed under modus ponens and that
for any *axiom* X of $\overline{K_4}$, the reasoner believes X and also believes that
he believes X. Will he necessarily believe all sentences that are
believed by all reasoners of type 4? (Remember, he may not be
normal!) The answer is given following Problem 2.

2

Now, substitute $\overline{G}$ in place of K_4. Will the reasoner above necessarily believe all sentences that are believed by all reasoners of
type G?

The answer to the above problems is given by a well-known
theorem about the modal systems K_4 and G (and which also applies
to $\overline{K_4}$ and $\overline{G}$), which we are about to state and whose proof we will
sketch.

242

Let M be any modal system whose only inference rules are modus ponens and necessitation. We let M' be that modal system whose axioms are the axioms of M together with all formulas BX, where X is an axiom of M, and whose *only* inference rule is modus ponens. It is obvious that everything provable in M' is provable in M (because for any axiom X of M, BX is also provable in M, and so all axioms of M' are provable in M), but in general it is not true that everything provable in M is provable in M'. However, we have the following interesting result:

Theorem 1. If all axioms of K_4 are provable in M, then it *is* true that everything provable in M is provable in M' (and hence the systems M and M' prove exactly the same formulas).

Can the reader see how to prove Theorem 1?
(*Hint:* First show that if X is provable in M, then BX is provable in M'. Do this by showing that for any proof $X_1, \ldots, X_n$ in M, all of the formulas $BX_1, \ldots, BX_n$ are successively provable in M'.)

More Detailed Proof. Since modus ponens is a rule of M' and all tautologies are among the axioms of M', then M' is of course of type 1. Also the following three conditions hold for M':

(1) If BX and B(X⊃Y) are provable in M', so is BY—because B(X⊃Y)⊃(BX⊃BY) is an axiom of M' and M' is of type 1.

(2) If BX is provable in M', so is BBX—because BX⊃BBX is an axiom of M' and M' is of type 1.

(3) If X is an axiom of M, then BX is provable in M'—because it is even an axiom of M'.

Now, suppose a sequence $X_1, \ldots, X_n$ of formulas constitutes a proof in M. Each line of the proof either comes from two earlier lines by modus ponens, or from one earlier line by the necessitation rule, or is itself an axiom of M. Using facts (1), (2), and (3) above, it easily follows that BX_1 must be provable in M', then that BX_2 is provable in M', then that BX_3 is provable in M', and so forth down to BX_n. We leave the verification of this to the reader.

Now we know that if X is provable in M, then BX is provable in M'. Therefore, if X is provable in M', then BX is provable in M' (because if X is provable in M', it is also provable in M). And so we see that M' is normal (even though the necessitation rule is not initially given for M'). Then, given any proof $X_1, \ldots, X_n$ in M, it is easy to see that each of the lines $X_1, \ldots, X_n$ *itself* can be successively proved in M'. (We leave the verification of this to the reader.)

Corollary. The provable formulas of K_4 and K_4' are the same. The provable formulas of G and G' are the same.

The following theorem and its corollary can be proved by a similar argument.

Theorem 1. For any *sentential* modal system $\overline{M}$ in which all axioms of $\overline{K_4}$ are provable, and in which modus ponens and necessitation are the only inference rules, the provable sentences of $\overline{M}$ are the same as the provable sentences of $\overline{M}'$.

Corollary. The provable sentences of $\overline{K_4}$ are the same as those of $\overline{K_4}'$. The provable sentences of $\overline{G}$ are the same as those of $\overline{G}'$.

Of course the above corollary gives an affirmative answer to Problems 1 and 2.

We see by virtue of the corollary to Theorem 1 that the modal systems K_4 and G can be alternatively axiomatized using systems whose *only* inference rule is modus ponens.

· 29 ·

Some Strange Reasoners!

REASONERS WHO ARE ALMOST OF TYPE G

We shall say that a reasoner is *almost* of type G if he believes all sentences believed by all reasoners of type G (or, what is the same thing, if he believes all sentences provable in the modal system $\overline{G}$) and if his beliefs are closed under modus ponens. What possibly keeps him from being a reasoner of type G is that he may not be normal.

As we will prove, a reasoner who is almost of type G, unlike a reasoner who is of type G, *can* believe that he is consistent without losing his consistency. But then he must suffer from another defect —he cannot be normal!

We will now look more closely into this.

1

Given a consistent reasoner who is almost of type G and who believes he is consistent, find a proposition p such that the reasoner believes p, but can never know that he believes p!

2

Any abnormal reasoner must fail to believe at least one true proposition, because there is a proposition p such that he believes p but fails to believe Bp, yet Bp is true (since he believes p). Therefore he fails to believe the true proposition Bp.

It then follows from the last problem that given any consistent reasoner who is almost of type G and believes he is consistent, there must be at least one true proposition that he fails to believe. More startling is the fact that there must be at least one *false* proposition that he does believe!

What false proposition must he believe?

Exercise 1. State whether the following is true or false: Every non-normal reasoner of type 1 is consistent.

Exercise 2. State whether the following is true or false: Every non-normal reasoner of type 1 who believes all axioms of $\overline{K_4}$ must believe at least one false proposition.

REASONERS WHO ARE OF TYPE G*

As we will see, a reasoner who is almost of type G can not only believe in his own consistency without necessarily being inconsistent; he can even believe in his own *accuracy* without necessarily being inconsistent.

By a reasoner of type G* we shall mean a reasoner who is *almost* of type G and who believes all sentences of the form BX⊃X (he believes in his own accuracy). In other words, a reasoner of type G* is a reasoner of type 1 who believes all sentences provable in G and believes all sentences of the form BX⊃X.

Such a reasoner must, of course, also believe the sentence B⊥⊃⊥, and since he is almost of type G, then what we have proved in the solution to Problems 1 and 2 also holds good for him. And so we have established Theorem 1.

Theorem 1. For any consistent reasoner of type G^*:

(a) He believes that he is consistent, but can never know that he believes he is consistent!

(b) He also believes the false proposition $B{\sim}B{\perp}{\supset}BB{\sim}B{\perp}$ (i.e., he wrongly believes: "If I ever believe that I am consistent, then I will believe that I believe that I am consistent.")

We remind the reader that the sentence $B{\sim}B{\perp}{\supset}BB{\sim}B{\perp}$ is false for a consistent reasoner of type G^*, since $B{\sim}B{\perp}$ is true (he believes he is consistent), but $BB{\sim}B{\perp}$ is false (he doesn't believe that he believes that he is consistent).

It of course follows from Theorem 1 that *any* reasoner of type G^* must have at least one false belief, because if he is consistent, then he does (by Theorem 1), and if he is inconsistent, he certainly does!

Minimal Reasoners of Type G^*. By the modal system G^* is meant the system whose axioms are all the provable formulas of G together with all formulas of the form $BX{\supset}X$, and whose only inference rule is modus ponens. We shall let $\overline{G^*}$ be the system G^* whose axioms are restricted to sentences—that is, the axioms of $\overline{G^*}$ are all sentences provable in $\overline{G}$, plus all *sentences* of the form $BX{\supset}X$. The only inference rule of $\overline{G^*}$ is modus ponens. (We can easily show by an argument similar to the one used in Chapter 27 that the provable sentences of $\overline{G^*}$ are the same as the provable *sentences* of G^*.) By a *minimal* reasoner of type G^* we shall mean a reasoner who believes those and only those sentences provable in $\overline{G^*}$. It is easy to show that all reasoners of type G^* must believe all sentences provable in $\overline{G^*}$ ("believe" here falls under the egocentric interpretation, of course). Therefore a reasoner is a minimal reasoner of type G^* if and only if he believes those and only those sentences that are believed by all reasoners of type G^*.

Since every reasoner of type G^* has at least one false belief, so does a minimal reasoner of type G^*. It hence follows that there is at least one sentence provable in $\overline{G^*}$ that is false for $\overline{G^*}$ (false, when "B" is interpreted as provability in $\overline{G^*}$). And so we have Theorem 2.

Theorem 2. The modal system $\overline{G^*}$ is *not* self-referentially correct.

In light of Theorem 2, the reader may well wonder how the modal system $\overline{G^*}$ could be of any use. Well, just because there is a provable sentence of $\overline{G^*}$ that is false *for* $\overline{G^*}$ does not mean that there isn't some other interpretation of "B" in which all provable sentences of $\overline{G^*}$ are true. Is there such an interpretation? Yes, there is—and a very important one.

Theorem 3. Every sentence provable in $\overline{G^*}$ is true *for the modal system $\overline{G}$.*

This means that every sentence provable in $\overline{G^*}$ is true *if "B" is interpreted as provability in $\overline{G}$, rather than in $\overline{G^*}$.*

3

Why is Theorem 3 correct?

Theorem 4 is an easy corollary of Theorem 3.

Theorem 4. The system $\overline{G^*}$ is consistent.

4

Why is Theorem 4 a corollary of Theorem 3?

Theorem 4, of course, implies that any *minimal* reasoner of type G^* is consistent. And so a minimal reasoner of type G^* is consistent, he believes he is consistent, but can never believe that he believes he is consistent (by Theorem 2). Stated alternatively in terms of the modal system $\overline{G^*}$: it is consistent, it can prove its own consistency, but can never prove that it can prove its own consistency! Also, the modal system $\overline{G}^*$ is not normal.

The Completeness of $\overline{G}^*$ for $\overline{G}$. We shall now state a further result whose proof unfortunately goes beyond the scope of this book.

Given two modal systems M_1 and M_2, we have defined M_1 to be *correct* for M_2 if every sentence provable in M_1 is *true* for M_2. Let us say that M_1 is *complete* for M_2 if every sentence that is true for M_2 is actually provable in M_1.

Theorem 3 says that the modal system $\overline{G}^*$ is *correct* for the modal system $\overline{G}$. Well, it also happens to be *complete* for $\overline{G}$—every sentence *true* for $\overline{G}$ is *provable* in $\overline{G}^*$. And so the provable sentences of $\overline{G}^*$ are precisely the sentences that are true for $\overline{G}$. Thus a sentence is provable in $\overline{G}^*$ if and only if it is *true* for all reasoners of type G.

REASONERS OF TYPE Q (QUEER REASONERS)

By a *queer* reasoner—or a reasoner of type Q—we shall mean a reasoner of type G who believes that he is *inconsistent.* Can a queer reasoner be consistent? We will soon see that he can! Of course, every queer reasoner is normal.

By the modal system $\overline{Q}$, we shall mean the modal system $\overline{G}$ with the sentence $B\bot$ added as an axiom. By a *minimal* reasoner of type Q, we mean a reasoner who believes those and only those sentences provable in the modal system $\overline{Q}$—or, what is the same thing, a reasoner who believes all and only those sentences believed by all reasoners of type Q.

Theorem 5. The modal system $\overline{Q}$ is not self-referentially correct, but it *is* consistent.

5

Why is Theorem 5 true?

It of course follows from Theorem 5 that any minimal reasoner of type Q is consistent, although he believes he isn't!

A comparison. It is amusing and instructive to compare minimal reasoners of types G, G*, and Q.

(1) A minimal reasoner of type G is consistent, but can never know it.

(2) A minimal reasoner of type G* is consistent, believes he is consistent, but can never know that he believes he is consistent.

(3) A minimal reasoner of type Q believes he is inconsistent, but he is wrong—he is actually consistent.

SOLUTIONS

1 · One such proposition p is the proposition that the reasoner is consistent!

We are given that he believes ~B⊥ and we are to show that if he is consistent, he cannot believe B~B⊥. Well, since he believes all sentences provable in G, he certainly believes all tautologies, and since his beliefs are closed under modus ponens, he is certainly of type 1. He believes ~B⊥, so he believes B⊥⊃⊥. If he believed B~B⊥, he would believe B(B⊥⊃⊥). However, he does believe B(B⊥⊃⊥)⊃B⊥ (because he believes all sentences provable in G). And so he would then believe B(B⊥⊃⊥) and believe B(B⊥⊃⊥)⊃B⊥, hence he would believe B⊥. But since he believes ~B⊥, he would be inconsistent.

This proves that if he believes B~B⊥, he would be inconsistent. But we are given that he is consistent, hence he can never believe B~B⊥ (even though B~B⊥ is true).

2 · B~B⊥ is true, but since he doesn't believe it, then BB~B⊥ is false —hence B~B⊥⊃BB~B⊥ is false. But he must believe this sentence (because it is of the form BX⊃BBX where X=~B⊥), hence it is an axiom of G. And so he believes the false sentence B~B⊥⊃BB~B⊥. (He wrongly believes: "If I should believe I am consistent, then I would believe that I believe that I am consistent." This belief is wrong, since in fact he *does* believe he is consistent, but *doesn't*— and never will—believe that he believes he is consistent.)

Incidentally, by the same argument, *any* nonnormal reasoner who believes all sentences provable in K_4 must have at least one false belief. There is some p such that he believes p but doesn't believe Bp, so Bp⊃BBp is false (for such a reasoner), yet it is an axiom of K_4, and the reasoner therefore believes it. This answers Exercise 2.

Exercise 1. It is true! If he were inconsistent and of type 1, he would believe *all* propositions, hence there would be no proposition p such that he believes p and doesn't believe Bp (because he believes both p and Bp, since he believes everything). Therefore every inconsistent reasoner of type 1 must be normal—or, put another way, every nonnormal reasoner of type 1 is consistent.

3 · We proved in the last chapter that G is self-referentially correct, hence:

(1) All sentences provable in $\overline{G}$ are true for $\overline{G}$.

Also:

(2) All sentences of the form BX⊃X are true for $\overline{G}$.

The reason for (2) is that if BX is true for $\overline{G}$, then X is provable in $\overline{G}$ (that's what it means for BX to be true for $\overline{G}$), and X must therefore be true for $\overline{G}$ (since $\overline{G}$ is self-referentially correct). And so, if BX is true for $\overline{G}$, so is X—which means that BX⊃X is true for $\overline{G}$.

By virtue of (1) and (2), every *axiom* of $\overline{G^*}$ is true for $\overline{G}$. Since the only inference rule of $\overline{G^*}$ is modus ponens, and since the set of sentences that is *true* for $\overline{G}$ is closed under modus ponens (if X and X⊃Y are true for $\overline{G}$, then obviously Y is true for $\overline{G}$), it follows that every sentence provable in $\overline{G^*}$ must be true for $\overline{G}$. Thus the system $\overline{G^*}$ is *correct* for $\overline{G}$.

4 · Since $\overline{G^*}$ is correct for $\overline{G}$, then if ⊥ were provable in $\overline{G^*}$, it would be true for $\overline{G}$, which is absurd. Therefore ⊥ is not provable in $\overline{G^*}$, and so $\overline{G^*}$ is consistent (even though it is not self-referentially correct).

5 · The system $\overline{Q}$ is of type G, and therefore by (c) of Lemma B, Chapter 27 (page 233), all axioms of $\overline{G}$ are true for $\overline{Q}$.

Now, let us momentarily assume that the sentence B⊥ is true for $\overline{Q}$. We then get the following contradiction: If B⊥ is true for $\overline{Q}$, then since all the other axioms of $\overline{Q}$ (i.e., the axioms of $\overline{G}$) are true for $\overline{Q}$, we would have *all* the axioms of $\overline{Q}$ being true for $\overline{Q}$. Then by Corollary A_1 of Lemma A, Chapter 27 (page 231), the system $\overline{Q}$ would be self-referentially correct. And so, if B⊥ is true for $\overline{Q}$, then $\overline{Q}$ is self-referentially correct. On the other hand, to say that B⊥ is true for $\overline{Q}$ is to say that ⊥ is provable in $\overline{Q}$, and since ⊥ is obviously false for $\overline{Q}$, this would mean that $\overline{Q}$ is *not* self-referentially correct. It is therefore contradictory to assume that B⊥ is true for $\overline{Q}$. Hence B⊥ is false for $\overline{Q}$, which means that ⊥ is *not* provable in $\overline{Q}$, and so $\overline{Q}$ must be consistent! But also B⊥ is an axiom of $\overline{Q}$, hence of course provable in $\overline{Q}$, and since it is false for $\overline{Q}$, then $\overline{Q}$ is not self-referentially correct. And so we see that $\overline{Q}$ is consistent but not self-referentially correct.

· 30 ·

In Retrospect

WE BEGAN this study with introspective reasoners and have wound up in the labyrinths of modal logic. Let us summarize some of the main things we have learned on the journey.

1. An accurate Gödelian system of type 1 cannot prove its own accuracy—i.e., it cannot prove all propositions of the form BX⊃X.

2. Any Gödelian system of type 1 that can prove its own accuracy is not only inaccurate, but peculiar—i.e., there must be a proposition p such that p and ~Bp are both provable.

3. Any Gödelian system of type 1* which can prove its own nonpeculiarity is peculiar.

4. (After Gödel's First Incompleteness Theorem.) Any normal, stable, consistent Gödelian system of type 1 must be incomplete. More specifically, if S is a normal system of type 1 and p is a proposition such that p≡~Bp is provable in S, then:

(a) If S is consistent, p is not provable in S.

(b) If S is consistent and stable, then ~p is also not provable in S.

5. (After Gödel's Second Theorem.) No consistent Gödelian system of type 4 can prove its own consistency.

6. A Gödelian system of type 4 can even prove that if it is consistent, then it cannot prove its own consistency—i.e., it can prove the proposition ~B⊥⊃~B(~B⊥).

7. (After Löb.) If S is a reflexive system of type 4, then for any

proposition p of the system, if Bp⊃p is provable in the system, so is p.

8. A system of type 4 is reflexive if and only if it is of type G.

9. A system of type 4 is Löbian if and only if it is of type G.

10. (After Kripke, de Jongh, Sambin.) Any system of type 3 in which all propositions of the form B(BX⊃X)⊃BX are provable must be of type 4 (and hence of type G).

11. A consistent system of type G cannot prove any proposition of the form ~BX—in particular, it cannot prove its own consistency.

12. A consistent and stable system of type G can neither prove its own consistency nor its own inconsistency.

13. (Semantical Soundness Theorems.) For any modal formula X, if X is provable in K, then it holds in all Kripke models; if it is provable in K_4, it holds in all transitive models; and if it is provable in G, then it holds in all transitive terminal models.

14. There do exist consistent stable systems of type G—for example, the machines of Fergusson and Craig, and the modal system $\overline{G}$. These systems, though consistent, cannot prove their own consistency.

15. The modal systems $\overline{K}$, $\overline{K}_4$, and $\overline{G}$ are not only consistent and stable, but are self-referentially correct. The same goes for the systems K, K_4, and G.

16. Neither of the systems $\overline{G^*}$ or $\overline{Q}$ is self-referentially correct, but both of them are consistent. The system $\overline{Q}$ is normal, but the system $\overline{G^*}$ is not.

17. (a) A minimal reasoner of type G is consistent, but can never know it.

(b) A minimal reasoner of type G^* is consistent and believes that he is consistent, but can never know that he believes he is consistent.

(c) A minimal reasoner of type Q believes he is inconsistent, but is really consistent.

This seems like a good stopping place. There are many more fascinating things about the modal system G. By far the best general reference currently in existence is Boolos's *The Unprovability of Consistency*, which I heartily recommend as a follow-up to this book. To whet the reader's appetite, let me give a sample of a beautiful result—a fixed-point theorem—whose proof can be found there.

Consider a modal formula with only one propositional variable, say the letter p. Let us denote such a formula as A(p). For any modal sentence S, by A(S) we mean the result of substituting S for every occurrence of p in A(p). For example, if A(p) is the formula p⊃Bp, then A(S) is the sentence BS⊃S. A sentence S is called a *fixed point* of A(p) if the sentence S≡A(S) is provable in G. In Problem 4 of Chapter 19, we asked the reader to find a proposition p such that any reasoner of type G will believe p≡~Bp, and we found ~B⊥ to be a solution. Thus every reasoner of type G will believe ~B⊥≡~B~B⊥, hence this sentence is provable in G. This means that ~B⊥ is a fixed point of the formula ~Bp. The formula B~p also has a fixed point—namely, B⊥, as we found in the solution to Problem 5, Chapter 19. The reader might find it a profitable exercise to try to show that the formula Bp⊃B⊥ has a fixed point—namely, BB⊥⊃B⊥.

Not every formula A(p) has a fixed point; for example, the formula ~p doesn't (otherwise the system G would be inconsistent, which we know is not so). Now, a formula A(p) is called *modalized* in p if every occurrence of p in A(p) lies in the part of A(p) of the form BX, where X is a formula. (Examples: Bp⊃BBp is modalized in p; Bp⊃p is not, but B(Bp⊃p) is.) The logicians Claudio Bernardi and C. Smorynski have independently proved that any formula A(p) that *is* modalized in p *does* have a fixed point S—moreover, the formula B(p≡A(p))⊃B(p≡S) is provable in G. This result is known as the Bernardi-Smorynski Fixed-Point Theorem.

Fixed points are remarkable things. By virtue of the self-referential correctness of the system G, any fixed point S of a formula A(p) is not only provable in G if and only if A(S) is provable, but is also *true* (for G) if and only if A(S) is true—because the provability of

S≡A(S) implies its truth. Let us say that a formula A(p) *applies* to a sentence S if A(S) is true (for G). A fixed point of a formula can then be thought of as a sentence that asserts that the formula applies to that very sentence.

More generally, consider a formula A(p,q) having no propositional variables other than p and q. For any formula X, by A(X,q) is meant the result of substituting X for p in A(p,q). By a fixed point of A(p,q) is meant a formula H having no variables other than q, such that the formula H≡A(H,q) is provable in G. The logicians D. H. J. de Jongh and Giovanni Sambin have proved that if A(p,q) is modalized in p (but not necessarily in q), then A(p,q) has a fixed point H; moreover, the formula B(p≡A(p,q))⊃B(p≡H) is provable in G.

Examples are already familiar from Chapter 19. Bq is a fixed point of B(p⊃q), by Problem 6, and Bq⊃q is a fixed point of Bp⊃q, by Problem 7. Indeed, reflexivity is equivalent to there being a fixed point for Bp⊃q. The remarkable thing is that for a modal system of type 4, the existence of a fixed point for the one formula Bp⊃q is enough to guarantee fixed points for *all* formulas A(p,q) that are modalized in p. A proof of this can also be found in Boolos.

CONCLUDING REMARKS

I hope I have given the reader some feeling for why these modal systems are so interesting. We have seen that they can be interpreted both internally (self-referentially) and externally (as applying to provability in other mathematical systems), as well as to reasoning processes—both for naturally intelligent beings (some humans and other animals) and for artificially intelligent mechanisms (such as computers). What applications this may have in the field of psychology is something that might be worth further investigation.

It is a happy turn of fate that the field of modal logic, which historically arose out of purely philosophical interests, should have turned out to be so important today in proof theory and computer science—this by virtue of the theorems of Gödel and Löb and of the work of those who have subsequently looked at proof theory from a modal-theoretical viewpoint. And now even those philosophers who in the past have taken a dim view of the significance of modal logic are forced to realize its mathematical importance.

The past philosophical opposition to modal logic has been grounded roughly in three quite different (and incompatible) beliefs: First, there are those who believe that everything true is necessarily true, and hence that there is no difference between truth and necessary truth. Second, there are those who believe that nothing is necessarily true, and hence that for any proposition p, the proposition Np (p is necessarily true) is simply false! And third, there are those who claim that the words "necessarily true" convey no meaning whatsoever. And so each of these philosophical types has rejected modal logic on his own grounds. Indeed, one well-known philosopher is reputed to have suggested that modern modal logic was conceived in sin. To which Boolos has aptly replied: "If modern modal logic was conceived in sin, then it has been redeemed through Gödliness."

A NOTE ON THE TYPE

The text of this book was set in Electra, a type face designed by William Addison Dwiggins (1880–1956) for the Mergenthaler Linotype Company and first made available in 1935. Electra cannot be classified as either "modern" or "old style." It is not based on any historical model, and hence does not echo any particular period or style of type design. It avoids the extreme contrast between thick and thin elements that marks most modern faces, and it is without eccentricities that catch the eye and interfere with reading. In general, Electra is a simple, readable type face that attempts to give a feeling of fluidity, power, and speed.

W. A. Dwiggins began an association with the Mergenthaler Linotype Company in 1929 and over the next 27 years designed a number of book types, including Metro, Electra, Caledonia, Eldorado, and Falcon.

Composed, printed and bound by The Haddon Craftsmen, Inc., Scranton, Pennsylvania

DESIGNED BY PETER A. ANDERSEN